"At the *Gazette-Mail*, Eyre's career has been the stuff of quiet legend. . . . For years, I have nursed a hope: that, someday, Eyre would write a book on what it is to do big work in a small newsroom, and what it is we'll lose if those newsrooms are no more. Now that day has come."
—The New Yorker

"Compelling and terrifying . . . In many ways [*Death in Mud Lick* is] as interesting as a John Grisham courtroom thriller, featuring an extended series of legal actions, the requisite heroes and villains, and personal problems that add to the drama."
—The Daily Beast

"Eric Eyre represents the absolute best of newspaper reporting. . . . *Death in Mud Lick* is a riveting, intimate look at the corporate greed, regulatory failure, and lobbying shenanigans that led to pill mills complete with 'courtesy snacks' and cash registers so full they wouldn't close."
—Beth Macy, author of *Dopesick: Dealers, Doctors, and the Drug Company That Addicted America*

D0111824

More praise for *Death in Mud Lick*

"*Death in Mud Lick* is full of sharp reporting and storytelling. But is also infused with the kind of compassion and moral outrage necessary to understand how a man-made disaster like this one happens."

—*Pittsburgh Current*

"Eyre mined for the truth not with a hard hat or heavy machinery, but with notepads, pens, and a doggedness to keep digging. *Death in Mud Lick* is for readers who root for the underdog, who want justice served, and who love newspapers."

—*The Roanoke Times*

"With searing storytelling and deep investigative reporting, Eric Eyre has written an indispensable book that you won't be able to put down."

—Anna Sale, host of the podcast
Death, Sex and Money

"Packed with colorful details and startling statistics, this page-turning journalistic thriller shines a brilliant spotlight on a national tragedy."

—*Publishers Weekly* (starred review)

"A riveting drama of crime, collusion, coverup . . . *Death in Mud Lick* attaches names, stories, and vivid characters to the major public-health story of our times."

—James Fallows, author of *Our Towns: A 100,000-Mile Journey into the Heart of America*

"At a time when real journalism is under attack, *Death in Mud Lick* stands as a clenched fist of rebuke. . . . Eyre's book is a thrilling recounting of how it all went down in the tradition of *Call Northside 777* and *Spotlight*."

—Brian Alexander, author of *Glass House: The 1% Economy and the Shattering of the All-American Town*

"Compellingly told, [*Death in Mud Lick*] adds another layer to the reporting of the opioid crisis . . . a tale of compassionate people deeply wronged and a dogged journalist who won't stand for it."

—*Booklist* (starred review)

"Heart-wrenching . . . engrossing reportage on an issue that can't receive too much attention."

—*Kirkus Reviews*

"How did they get away with it for so long, raking in billions of dollars even as thousands died from overdoses? Read *Death in Mud Lick* and understand. It is a stunning story, and Eric Eyre tells it with compassion, grit, deep knowledge, and the 'sustained outrage' (as he puts it) that is the rocket fuel of great journalism."

—Dan Fagin, author of the Pulitzer Prize–winning *Toms River*

"Timely and well documented, with appeal to a broad range of readers."

—*Library Journal* (starred review)

"A uniquely American tale and heartbreaking account that will leave every reader certain on how we got into this nightmare otherwise known as the opioid crisis."

—Ryan Hampton, nationally recognized recovery activist and author of *American Fix: Inside the Opioid Addiction Crisis—and How to End It*

DEATH IN MUD LICK

A Coal Country Fight
against the Drug Companies
That Delivered the Opioid Epidemic

Eric Eyre

SCRIBNER

New York   London   Toronto   Sydney   New Delhi

Scribner
An Imprint of Simon & Schuster, Inc.
1230 Avenue of the Americas
New York, NY 10020

First Scribner trade paperback edition April 2021

SCRIBNER and design are registered trademarks of The Gale Group, Inc.,
used under license by Simon & Schuster, Inc., the publisher of this work.

For information about special discounts for bulk purchases,
please contact Simon & Schuster Special Sales at 1-866-506-1949
or business@simonandschuster.com.

The Simon & Schuster Speakers Bureau can bring authors to your live event.
For more information or to book an event, contact the Simon & Schuster Speakers Bureau
at 1-866-248-3049 or visit our website at www.simonspeakers.com.

Manufactured in the United States of America

1   3   5   7   9   10   8   6   4   2

Library of Congress Cataloging-in-Publication Data has been applied for.

ISBN 978-1-9821-0531-0
ISBN 978-1-9821-0532-7 (pbk)
ISBN 978-1-9821-0533-4 (ebook)

To Lori and Toby

Pray for the dead, and fight like hell for the living.

—Mother Jones

Contents

CONTENTS

Preface

In two years, out-of-state drug companies shipped nearly 9 million opioid pain pills to Kermit, West Virginia, a town with 382 people. The quintessential coal town, split by a pair of railroad tracks, was the home of Sav-Rite Pharmacy, which once had the dubious distinction of being among the country's top sellers of a highly addictive prescription painkiller called hydrocodone—packaged under brand names such as Lortab and Vicodin. Sav-Rite was the only game in town. The pharmacy's owner, Jim Wooley, sold used cars on the side, right there in the gravel lot beside Sav-Rite. It was quite a racket.

Kermit didn't have nearly enough customers to buy that many pain pills. You could step into just about any "pain management clinic" in the county and walk out with a bogus prescription for $150. Wooley—he pronounced it "OO-LEE"—had established a considerable footprint. Sav-Rite's clientele would travel from hundreds of miles away, from Ohio, North Carolina, Tennessee, and even Florida. Word spread fast and far when a pharmacy would fill any prescription so long as you paid in cash. When folks started asking questions about Wooley's booming business in the middle of nowhere, he had a ready answer. His customers were mostly tourists, just passing through Kermit, on their way to hunt or fish or ride four-wheelers in the mountains. But it was Sav-Rite that had become the tourist destination. Cars and pickups were backed up, trying to squeeze into the drive-through lane, choking Highway 52 through town. Wooley was a salesman, through and through, and he recognized that a waiting customer wasn't a happy customer. So he dragged a camping trailer onto the parking lot and sold hot dogs and chips and soda pop out of it. The concessions were cheap, the customers were happy, and Wooley could make a few extra bucks outside the pharmacy to couple with the millions he was making inside selling opioids. To the tourists.

To keep pace with demand, he needed reliable suppliers. There was no shortage. One was McKesson Corporation. It ranked sixth in the Fortune 500. A couple of years back, McKesson's CEO was the highest-paid corporate executive in the land. And the company didn't hesitate to fill Jim Wooley's round-the-clock orders. In 2006 and 2007, McKesson shipped 5 million hydrocodone pills to Sav-Rite, no questions asked. The following year, when Wooley's actions started raising suspicions—he opened a sham pain clinic up the road where addicts would pick up rubber-stamped prescriptions that only Sav-Rite would honor—McKesson, like a good corporate citizen, cut the pharmacy off. For two years. Once the authorities stopped snooping around, however, the global drug distributor resumed deliveries of hydrocodone and other powerful pain medications to Sav-Rite. But then Wooley got arrested for filling bogus prescriptions, and, well, that terminated the business relationship for good. It was lucrative while it lasted. McKesson's CEO denied responsibility and faced no penalty. Wooley almost got off scot-free as well. Prosecutors recommended no prison time for the pharmacist-turned-entrepreneur.

Across West Virginia, other small towns like Kermit were also drowning in prescription painkillers. Thirty miles east, McKesson combined with wholesale drug giant Cardinal Health—the fourteenth-largest US company—and two regional distributors to deliver 16.6 million pain pills over a decade to a single drugstore in Mount Gay, which has all of seventeen hundred residents. Those same companies, along with AmerisourceBergen—ranked twelfth in the Fortune 500—shipped 20.8 million prescription opioids to two pharmacies four blocks apart in Williamson, a town with twenty-nine hundred people, and only twenty miles from Kermit. Williamson was so overrun with painkillers that the locals started calling it Pilliamson. The white coats and blue suits made a fortune.

This was unbridled profiteering, yes, and it came with an undeniable public health cost. The pills were lethal. Take too many all at once, and you stopped breathing. People were taking hydrocodone and its more powerful cousin OxyContin, and they were accidentally overdosing in record numbers. Mingo County, where Kermit and Williamson are located, had one of the highest overdose death rates in the nation, according to the Centers for Disease Control. As the addiction crisis spread across the country, some health advocates sounded the alarm, but industry lobbyists snuffed out policymakers' efforts to stop the scourge. They found politicians willing to do their bidding. The regulators—the DEA, the pharmacy board—failed to do their jobs. Pablo Escobar and El Chapo couldn't have set things up any better. So the pills kept flowing, the number of deaths mounting. Federal laws and court orders kept the companies' dark secrets hidden from the public. They left nothing to chance. It was all too big. And, truth be told, they almost got away with it, the biggest heist amid one of the biggest public health crises in US history. Almost.

But there was something the corporate pill peddlers didn't forecast, something that took them by surprise: an unlikely alliance between an ex-con and the crusading lawyer who couldn't keep her out of jail. Starting in 2007, they slung accusatory stones up, up, up the drug supply chain, from doctors to pharmacists to drug distributors.

As a statehouse reporter with the *Charleston Gazette-Mail*, I stumbled into the middle of their legal battle in 2013, uncovering secrets and lies that set up a collision course with three of America's largest corporations. That summer, I received a tip that Cardinal Health had helped pay for the inaugural party of West Virginia's newly elected attorney general, Patrick Morrisey. Cardinal's lawyer had headed Morrisey's campaign transition team, and Morrisey's wife had lobbied for Cardinal in Washington, DC, pocketing millions of dollars for her K Street firm. The previous attorney general—a twenty-year incumbent—had sued Cardinal on behalf of the citizens of West Virginia. Now, Morrisey, after Cardinal's top executives helped bankroll his campaign, was overseeing the suit; lawyers close to the case contended he was trying to sabotage it. Morrisey insisted he had stepped aside from the lawsuit, but I unearthed letters showing he had met privately with Cardinal lawyers about it, and court documents and emails revealed he was giving staff "specific instructions" on how to handle the suit. In retaliation, Morrisey set out to derail my investigation with one of his own—against my employer, a tenacious small newspaper in financial peril. His benefactors were counting on him to slam shut the door. But after the paper successfully fought to unseal court documents that the drug distributors wanted to hide from the public, the attorney general handed over previously confidential records that showed the companies' insidious pursuit of profits. Along the way, I wrote hundreds of stories about the devastation and misery that opioids had inflicted upon our state. I kept digging for answers, the smaller articles snowballing into the larger story of how it happened, how drug companies flooded small towns with millions of prescription opioids, and how they got caught. It all began with a seemingly unremarkable death in a place called Mud Lick.

PART I

1

A Death in Mud Lick

At sunup, Debbie Preece drove north on the two-lane blacktop that traced the Tug Fork of the Big Sandy River, hurtling onto a rutted gravel road that tunneled deep into the woods. She stopped with a jolt at the rust-bitten trailer in Mud Lick. The coroner had already picked up her brother's body and transported it to the morgue for autopsy. Debbie insisted that someone show her where William "Bull" Preece had spent his last hours. She was directed to a back bedroom, vacant save for a dresser and a torn mattress set atop a box spring. The sheriff's deputies had already removed the blood-spattered clothes and swept up the residue of crushed pills.

It was the first Monday in October 2005, five years since Bull had fallen from a ladder and injured his back at Penn Coal mine and secured that first prescription for pain pills, which led to another and another. Bull kept finding doctors to prescribe OxyContin and Lortab. He had been taking painkillers for two years before Debbie realized he was addicted. He'd slump into the sofa at her house and fall asleep. She tried to reason with him. She checked him into rehab. He'd stay a few days, then relapse. He lost his home after failing to pay the mortgage. He became a stranger to Debbie. The contents of a pill bottle would transform a once strong and proud coal miner into a coward prowling darkened streets and dive bars for his next fix. He tried to shed the addiction at a methadone clinic. He stopped going after six months. He told his sister he was taking pain medication, but he assured her he wasn't snorting pills or shooting up.

"Be careful," she had told him as he was leaving their mother's house in Kermit, the last time they spoke.

"See you tomorrow," he said.

After rifling through the single-wide, Debbie stepped outside, her platinum-blond hair afire in the morning sun, her brown eyes, rimmed with red, narrowing to scan the depths of the hollow. At forty-eight, she was three years older than Bull, so nicknamed because he was a bullheaded child.

Debbie noticed a white truck parked beside a shotgun-frame house with a broken porch. It was Bull's Ford Explorer. She swung open the passenger-side door and retrieved a stack of family photographs—Bull with his thirteen brothers and sisters, Bull with his arm around his father, Bull dressed in a navy firefighter's uniform and white cap beside the fire truck outside the Kermit VFD.

A man emerged from the house and waved Debbie inside. He wanted to show her something he had discovered in the truck's glove box. Prescription receipts and four orange plastic prescription bottles. They were empty. Every one of them.

The evening before his death, Bull and his estranged wife and another man had been drinking beer at a honky-tonk called Sweeties Teardrop Inn. They drove back to Mud Lick at 1:00 a.m. that Sunday. Bull fell asleep on the mattress. The man swiped Bull's truck and drove across the Tug Fork into Kentucky to buy more beer. On the way back, he lost control of the Explorer, ran it up a hill, and slammed into the neighbor's porch. Someone rushed the man who wrecked Bull's truck to the hospital in Louisa, Kentucky. He slipped out of the emergency room before the authorities figured things out and tried to arrest him. He got back to the trailer around 1:00 p.m., fell onto a couch, and turned on the television to watch the Cincinnati Bengals game. Bull was still on the mattress in the bedroom. He hadn't moved. Nobody bothered to wake him.

The recounting of events was unsettling. Debbie rolled the hard plastic bottles in her palm, then gripped them tight. The white labels revealed the missing contents: ninety Valium; sixty oxycodone; ninety OxyContin, an extended-release form of oxycodone; and thirty Zestril tablets. She recognized the drugs. A sedative, painkillers, a blood-pressure medication. She knew the dangers of mixing them. She had attended too many funerals not to know.

She squinted at the labels more closely, the fine print displaying an eleven-digit code, as well as the names of a pharmacy she had frequented dozens of times and a doctor she had known for years. Dr. Donald Kiser had worked at the Wellness Center and the hospital emergency room in Williamson, the Mingo County seat, a twenty-minute drive from her home in Kermit. But he had written Bull's prescription from his new office in Marietta, Ohio, three hours away. Nothing kept Kiser from practicing in Ohio, even though his lies had cost him his license in West Virginia. In late February 2005, Mingo County deputies had arrested Kiser and charged him with trading prescriptions for sex. The medical board rejected his request for a new license after Kiser marked no to a question about whether he had been charged with a crime during the past two years. Kiser alleged that Mingo authorities had trumped up allegations against him in retaliation for a lawsuit he had filed against them near the end of the previous year. That suit claimed Mingo deputies had falsely arrested him during a child custody dispute with his ex-wife. Debbie didn't trust Mingo County authorities either. She didn't discount that Kiser might have been railroaded. She had been willing to give him the benefit of the doubt.

The move to Ohio didn't slow down Kiser's business. His Mingo County clientele wouldn't desert him. Every week or so, a van would pick up Bull and other passengers at stops along the two-lane and shuttle them north to Marietta. The fare was paid in pills—twenty for each prescription filled from each passenger. The drivers, in turn, sold the painkillers on the black market. Kiser arranged the shuttle service. Everyone made a lot of money.

Thirteen days before seeing Kiser, Bull had picked up a prescription for ninety hydrocodone pills for pain and sixty Xanax for anxiety at a recently opened pain clinic in Stonecoal, just north of Kermit, and he had receipts for another 120 hydrocodones and ninety Xanax prescribed there that month by a doctor who had never laid eyes on Bull. That, added to Kiser's prescriptions, placed six hundred and thirty pills in his hands over the past forty-five days, or nearly three times the recommended dosage for a patient with severe pain—all dispensed by Sav-Rite Pharmacy, the lone drugstore in Kermit. He paid $558 in cash.

Bull's last prescription was filled on September 29. He'd lumbered into the Sav-Rite on Lincoln Street, and the pharmacy's owner was there that day. His name was on the storefront sign: JIM WOOLEY'S SAV-RITE PHARMACY. He was sixty-eight now, his beard bushy and white. Wooley was affable, friendly to his customers, always a smile on his face, a people person, a salesman. Bull's prescription caught Wooley's attention, but not because it was written by Dr. Kiser at a clinic 170 miles away in Marietta. Wooley not only knew Kiser, but he had recently loaned him $5,000, two days after Kiser was arrested at the pain clinic in Williamson. Everybody knew one another in Mingo County. Bloody Mingo, the locals called it—birthplace of the Hatfield and McCoy feud, the Matewan massacre, a place where mine owners and union workers settled disputes with rifles, but now a place where drug merchants were calling the shots. The pain-addled addicts didn't stand a chance.

Wooley would claim at a deposition that he didn't know about Kiser's troubles with the Mingo County Prosecuting Attorney's Office and the West Virginia Board of Osteopathic Medicine. Nobody had told Wooley, and, no, he hadn't read about it in the *Williamson Daily News* or *Mountain Citizen*, the newspaper published across the river in Kentucky. What gave Wooley pause about Bull's prescriptions was the switch from hydrocodone to oxycodone. So he pulled Bull aside and counseled him, right there at the pharmacy, warned him about the change to a stronger painkiller, and not to take the extended and immediate release at the same time, and, whatever you do, don't chew on the OxyContin. That was his advice. That was it.

Three days later, Bull was dead. He was forty-five. The autopsy report showed he died of oxycodone intoxication. He had five times the lethal limit for the drug in his blood. His death was ruled an accident. Four days before Christmas, the state's investigation and postmortem for William H. Preece—Case No. WV 2005-1018—was officially closed.

Debbie took the empty prescription bottles home with her and secreted them in an upstairs dresser drawer for safekeeping. She wanted something to hold on to. Bull's death wouldn't be forgotten. He wouldn't be another number, a statistic, in the overdose death toll. Somebody was going to pay, no matter the repercussions, no matter what her enemies and the scandalmongers might dredge up about the past. There would be a reckoning.

Several weeks after they lowered Bull into the ground, the phone rang at Debbie's house. Dr. Kiser wanted to talk. It was urgent. Was he calling to offer his sympathies? No. He asked Debbie for a copy of Bull's MRI, the one taken after the ladder fall. Could she send it? The doctor needed something to put in Bull's empty patient file.

Some things about Bull's death still troubled Debbie. The bloody clothing, for instance. Could someone have murdered her brother, then tried to cover up the crime by making it look like an overdose? Debbie hadn't ruled that out. Stranger things had happened in Mingo County. She wanted an appointment to speak to the medical examiner. She wanted to know why the autopsy was taking so long. She kept calling the morgue.

"I want to look at the man who did the autopsy on my brother," Debbie told the secretary who answered the phone.

After finally receiving a copy of the autopsy report, Debbie hopped in her car and headed north on the four-lane highway to the medical examiner's office in Charleston. Corridor G, as the route was known, was built for trucks to haul coal out. Debbie wound her way through humps of mountains, their ridges glowing red, orange, and yellow with the change of season. She passed three prisons and two Walmarts during the hour-and-forty-five-minute trip. Her destination, a tan-painted building with barred windows, stood across the street from a NAPA Auto Parts store and the First Advent Christian Church. A receptionist directed Debbie and her questions to Dr. Zia Sabet, the chief deputy at the morgue, the man who conducted Bull's autopsy. Had he kept blood and DNA samples? Yes. Were there any marks on the body? No. Could her brother have been killed? No marks, no bruises, nothing to suggest someone had attacked Bull, Sabet said. But what about the bloody clothing beside the mattress in the trailer? She had seen evidence photos taken by the sheriff. Sabet told her he did not have the clothing.

Debbie had questioned everything about her brother's death. His Explorer had been taken for a joyride. Someone had stolen Bull's gold necklace and the pills left in his prescription bottles after he died. Why did his so-called friends let Bull lie passed out in the trailer so long? Why did it take them so long to call the sheriff? She was told someone hid her brother's pill bottles in a stack of cinder blocks while the ambulance made its way to the trailer. They later partied with Bull's remaining pain medications, she heard.

"I suspicioned everything when those things were missing," Debbie said.

"Your brother died of an overdose," Sabet told her.

There was also the pill residue—and an empty bottle of Lortab that belonged to the trailer's owner.

"My brother did not snort pills. When you crush them, that's what you crush them for."

"How do you know your brother didn't do that?" Sabet asked.

"Because he told me he didn't."

And she believed Bull. He had always been a man of his word, hooked but honest.

After Bull's funeral, Dr. Kiser was frantic. For weeks that autumn of 2005, he kept calling Debbie, asking whether her dead brother had ever had an MRI and offering to pay for the film. In fact, Bull did have one. It was in a big brown envelope stored upstairs beside the prescription bottles. But why on earth would the doctor need the scan? Kiser had an explanation: He admitted to Debbie that he had no justifiable reason for prescribing OxyContin to Bull, and the authorities—Kiser wouldn't identify them by name or agency—were starting to ask questions about Bull's death.

Debbie was helping to care for an elderly man on kidney dialysis, and she still thought Kiser was a respectable doctor. She would call him for instructions about what color of dialysis bag to administer based on the man's blood work. Balancing the chemical levels wasn't easy. She respected Kiser's medical advice and appreciated his help. He didn't ask for anything in return. Even when Debbie heard about the charges against Kiser in February 2005, the news didn't change her opinion.

"They're saying they're going to indict me over your brother's death," Kiser told Debbie.

Kiser was hatching a cover-up for Bull's death, and somehow he had convinced Debbie to take part. At first, Debbie thought she was doing the right thing. She wanted to help Kiser. He had always helped her. They kept talking on the phone, never in person. Then one Sunday in November 2005, Kiser's girlfriend showed up at Debbie's house. Debbie handed her the brown envelope with Bull's MRI tucked inside. She never looked at it. She couldn't remember where or when it had been taken—probably at the hospital emergency room in Louisa, after the mine accident.

And now that her brother was dead, she was giving it up, just like that, a willing accomplice. She watched Kiser's girlfriend—Debbie's former sister-in-law—drop the envelope in the trunk of her car, close the hood, and drive away to Kiser's new office in Ohio.

Weeks later, Debbie started having second thoughts. Everyone knew Kiser was writing prescriptions for pain pills. Why else would vanloads of addicts travel three hours to see him? She was helping the doctor who had written the prescription that killed her brother. Had she lost her mind? Had she forgotten her promise? Finally, she picked up the phone and called the Mingo County sheriff and then the prosecutor.

"I've done something I probably shouldn't have done," she confessed.

2

Prescription for Pain

In 2005, Americans such as Bull Preece were dying of drug over-doses in record numbers, and the DEA's new enforcement chief, Joseph Rannazzisi, was determined to do something about it.

Fatal overdoses had doubled in six years, and most of the thirty thousand deaths were being linked to prescription painkill-ers, despite repeated assurances from the pharmaceutical indus-try that patients wouldn't become addicted. In West Virginia, OxyContin was initially the pain medication of choice, then later Lortab. Both were being prescribed at unprecedented rates. Both were potential killers. Between 1999 and 2004, the number of West Virginians who lost their lives to accidental drug overdoses jumped 550 percent. The Mountain State had the highest over-dose death rate in the nation.

Rannazzisi, a twenty-year DEA agent who had risen up the ranks to head the agency's Office of Diversion Control, realized that traditional policing of doctors and pharmacists wasn't work-ing—it wouldn't slow the avalanche of opioids. Instead, Rannaz-zisi and the DEA began targeting prescription drug distributors, which could act as a choke point in the pharmaceutical supply chain. The wholesale distributors were the middlemen. They purchased prescription pills from drug manufacturers, hauled them to regional warehouses, then delivered the medications to

11

hospitals and pharmacies. The distributors were buyers and suppliers.

The DEA hadn't paid much attention to distributors before, and that suited the companies just fine. Then, in 2005, Rannazzisi came knocking. The gruff, sharp-tongued DEA agent arranged a series of one-on-one meetings with the nation's largest distributors—McKesson, Cardinal Health, and AmerisourceBergen. The three companies controlled 85 percent of drug shipments. They had the buying power to put a stranglehold on painkiller mania; Americans were consuming more than 80 percent of the world's supply of oxycodone, and 99 percent of its hydrocodone. The DEA pitched the "Distributor Initiative" as an education campaign. Rannazzisi showed the companies examples of their own customers whose ordering habits and patterns suggested the painkillers were winding up in the wrong hands. Moreover, the DEA sent three letters to distributors over the next two years, outlining their responsibilities under the Controlled Substances Act, a federal law that had been on the books for thirty years. Aimed to create a tightly regulated distribution system that supplied drugs to patients who genuinely needed them, the act classified prescription opioids for their high potential for abuse. The distributors kept them locked up in vaults at their warehouses.

The face-to-face meetings and letters reminded the distributors that they were obligated to flag and report to the DEA "suspicious" orders—instances when pharmacies' painkiller purchases abruptly spiked from one month to the next. Rannazzisi warned the execs that their companies' licenses could be revoked if they didn't abide by the law. The DEA followed up with five national conferences for distributors about prescription drugs and how to stop them from being diverted and resold illegally by dealers on the street. The companies had been put on notice.

The distributors didn't exactly warm to the new program. They were logistics managers, not criminal investigators. They weren't going to evaluate the veracity of prescriptions. They weren't going to block orders. They were making billions of dollars a year selling opioids. The DEA was being unfair and unreasonable, shoving unfunded mandates down their throats. But the distributors didn't push back. Not yet.

A decade earlier, the makers of OxyContin had unleashed a sales force that took the country by storm. Purdue Pharma representatives descended like locusts. They swarmed medical offices, pushing doctors to prescribe a new wonder drug that vanquished pain, everything from sore backs to toothaches. Their mission was to sell more Oxy.

Purdue, based in Stamford, Connecticut, armed its sales reps with sophisticated intelligence about doctors' prescribing practices. The drugmaker bought the information from a research firm that collected the data from pharmacies. When Purdue's reps walked in the door, they already knew the brands of pain medications and number of doses a doctor was prescribing, and even how patients paid for their prescriptions. The company preyed on family doctors, sending salesmen directly into their offices. Pain-clinic physicians such as Donald Kiser were the easiest marks. The sales reps knew far more about the drug's effects than the prescribers. Primary-care physicians were busy. They were more likely to take a Purdue rep's word that it was a disservice to patients, borderline malpractice even, to withhold OxyContin.

Purdue Pharma quickly figured out Appalachia was a ripe market for the drug. The region had a history of heavy painkiller use. Its workers mined coal, operated heavy machinery, putting them more at risk for accidents and injury. There was also isolation, poverty, despair. The little round pill could fix just about anything.

In West Virginia, Purdue deployed eight to ten sales reps—more than in other similar-size markets—on any given day to pitch Oxy. Their paychecks were among the best in the business.

The regional manager for the territory that covered the southern coalfields made $238,000 a year, according to a report in the *Lexington Herald-Leader*. Field reps were paid $150,000 annually, a large chunk of that in bonuses based on sales. Nationwide, Purdue was spending $150 million a year on its field reps. By 2000, OxyContin sales topped $1 billion. It was the world's bestselling painkiller. Then people started dying, first dozens, then hundreds, and Purdue and its aggressive sales strategy was fielding much of the blame.

In 2001, West Virginia became the first state to sue Purdue Pharma. Attorney General Darrell McGraw accused the company of coercive and deceptive practices that gave it an illegal monopoly in the pain-medication market. Paul Nusbaum, West Virginia's health secretary, told reporters, "When you have rural physicians who are isolated and get visits from drug reps, they are told that if they don't use OxyContin that they can be sued for malpractice."

McGraw's suit, filed in McDowell County, the state's poorest and most southern county, alleged that Purdue courted doctors by paying their travel costs to attend company-sponsored pain-management seminars at hotels across the country. In turn, the physicians wrote excessive OxyContin prescriptions. Doctors were even prescribing the drug for migraines and arthritis pain. Purdue called McGraw's allegations "completely baseless." Its statement said, "We want the many thousands of patients in West Virginia receiving pain relief from OxyContin tablets to rest assured that nothing in this case or any other case will cause us to abandon them or deter us from making sure that our drug is available to them."

By 2004, with the case set to go to trial, Purdue backtracked from that promise. A company lawyer filed a memorandum informing McDowell County circuit judge Booker T. Stephens that if jurors sided with McGraw and the state, the company would likely stop selling OxyContin in West Virginia. But Purdue didn't stop selling its painkiller in West Virginia or anywhere else. Just days into jury selection, Purdue agreed to pay $10 million to settle the case. It admitted no wrongdoing.

The settlement dogged McGraw for years. Instead of being celebrated as the first attorney general to take on Purdue Pharma and win, he was castigated for how he distributed the proceeds. A conservative political group, West Virginia Citizens Against Lawsuit Abuse, or CALA, led a campaign to discredit McGraw. Funded by the US Chamber of Commerce, it advocated for changes to the courts that would weaken consumer protection laws, making it harder to win lawsuits against corporations. CALA leaders blasted McGraw for using a third of the settlement funds to pay outside lawyers who helped his office with the case, even though that was the agreement from the start. The lawyers risked not being paid a dime if the state lost the case. They also wouldn't have been able to recoup the $400,000 they spent in litigation costs. McGraw shared the rest of the settlement money with state agencies, the University of Charleston's pharmacy school, and eight community corrections programs. He was accused of using the settlement to create a personal slush fund. Many West Virginians bought the narrative, and his Republican opponents in subsequent elections didn't hesitate to use it against him. McGraw's top aide said, "The national Chamber of Commerce and Citizens Against Lawsuit Abuse have but one purpose: to take on any elected officials who regulate companies that are Chamber/CALA members, and to replace those elected officials with individuals who will not enforce consumer credit and protection laws."

After the settlement, Purdue Pharma's problems didn't go away. They got worse. In May 2007, the OxyContin manufacturer and three of its executives pleaded guilty to federal charges of misleading the public about the risk of addiction from the drug. The company agreed to pay a $634.5 million fine for falsely claiming that its drug was less subject to abuse than other painkillers. Federal prosecutors in Roanoke, Virginia, not far from the state's border with West Virginia, declared that their victory had put pharmaceutical companies on notice that they wouldn't be able to get away with breaking the law to make a profit. A Justice Department administrator warned, "The things they plot in their boardrooms, the things they do behind closed doors, will not stay behind closed doors."

The federal case against Purdue made McGraw's settlement look like peanuts, that he'd been taken for a ride. It provided more fodder for his enemies.

The day before federal prosecutors announced their unprecedented plea deal with Purdue Pharma, a compliance officer at San Francisco–based McKesson combed through pain pill distribution numbers, searching a nationwide database to see whether any pharmacy customers were exceeding caps on drug purchases. Joe Rannazzisi's warnings had seemingly spurred McKesson to start a Lifestyle Drug Monitoring Program designed to screen shipments. The company decided to limit its US pharmacy customers to eight thousand hydrocodone and oxycodone pills per month (it wouldn't disclose to the DEA how it came up with that number), but the review of April sales orders turned up something startling: The regional warehouse in Washington Court House, Ohio, was shipping both Family Discount Pharmacy and Sav-Rite Pharmacy in West Virginia more than eight thousand prescription painkillers on average *every day*. And the two pharmacies were only thirty miles apart.

The compliance officer alerted a supervisor at McKesson's distribution center in Ohio. "I have been going through the daily dosage for all DCs [distribution centers]," he wrote in an email. "Two of your customers really jumped out at me—Family Discount Phcy and Sav-Rite Phcy. We need to document those ASAP, and I would like to understand their business that would drive the numbers."

16

Sav-Rite and Family Discount ranked among McKesson's top customers in West Virginia, even though it delivered drugs to much-larger chain pharmacies in cities throughout the state. Sav-Rite was buying more than thirty-six times McKesson's self-imposed limit. But that wasn't what it was reporting to the DEA. That June, McKesson represented that it had spent three months reviewing all its customer pharmacies, including Sav-Rite. The distributor assured the DEA that the pharmacies' drug purchases were "reasonable." It didn't mention the shipments flagged by its compliance officer, or that the problem in West Virginia required attention "ASAP." It didn't share the May 9 email with law enforcement.

Instead, the company put its trust in Jim Wooley. McKesson had Wooley fill out a form titled Declaration of Controlled Substance Purchases. The standardized form required pharmacies to explain their reasons for buying prescription painkillers and antianxiety medications. Wooley wrote that Sav-Rite "fills only controlled drugs from legitimate (licensed) physicians for only patients the pharmacy has a patient-pharmacy relationship. We do not"—and he underlined *not* twice—"do computer-internet prescription filling." That was a useful reassurance, since the DEA was investigating Florida–based pharmacies that bought pain pills from McKesson and sold them to online shoppers with phony prescriptions. Rannazzisi had also sought to revoke the drug-distribution giant's license to deliver pain meds from its warehouses in Lakeland, Florida, and Landover, Maryland, but the Ohio warehouse that shipped to Sav-Rite and other southern West Virginia pharmacies had so far escaped scrutiny. Wooley signed McKesson's declaration form on November 1, 2007. He had given his word. That was enough. That was McKesson's due diligence. Two weeks later, the company dispatched an employee to Kermit to talk to Wooley and tour the pharmacy. McKesson had never before sent someone to Sav-Rite—other than a truck driver to deliver prescription drugs. McKesson's inspector quickly figured out what was going on, but it was too late. The prescription opioids had already been sold, with most, presumably, diverted to the black market—over 2 million pills, at a time when unintentional overdoses were increasing faster in West Virginia than in any other state.

3

Kings of Kermit

In the summer of 2007, Debbie Preece tore open a letter from a law firm in Charleston. They wanted plaintiffs for a class action lawsuit against OxyContin maker Purdue Pharma. She'd been looking for a way to strike back. But when she called, it did not go well. The woman who answered, Debbie recalled, told her not to expect much from a potential settlement. Your brother was an addict, the woman said, a drug abuser. That was not how Debbie wanted Bull remembered.

"To you, his life wasn't valued at anything," Debbie told the woman. "But to me his life is priceless, bitch."

She hung up, then dialed another lawyer—Jim Cagle. They hadn't spoken in years. She'd hired him to keep her out of jail, back when the Preeces were the kings of Kermit.

The Preece dynasty started with Debbie's father, Wilburn "Wig" Preece, and his bootlegging and vote buying with moonshine, then the after-hours liquor sales. As the years passed, the family figured more money was in marijuana, PCP, and pills—Valium and Tylenol with codeine were top sellers in the 1980s. With cash pouring in, the Preeces purchased Cadillacs and Corvettes, speedboats, diamond rings, Nautilus equipment, and tanning beds. Debbie was married to the police chief. She served on the town council and ran the ambulance service. In Kermit, townsfolk called Debbie and her husband "J.R. and Sue Ellen," after characters on the TV show *Dallas*, though she more closely resembled J.R.'s niece Lucy Ewing, what with her petite, curvy figure, blond hair, and penchant for trouble.

The Preeces' activities were the worst-kept secret in Mingo County. The family sold drugs out of a trailer parked beside the town hall and police station. They posted a sign when inventory dwindled: OUT OF DRUGS. BE BACK IN 15 MINUTES. But more drugs were always to be had, a constant supply—the Preeces made sure of that. They had a railroad worker on the payroll. When the train rolled through town, he would toss bags of marijuana from the engine, the train never stopping, bag after bag landing beside the Norfolk Southern tracks, where they would be swept up in no time and deposited at the trailer beside the police station.

Wig Preece had bought off the sheriff, some of the county commissioners, some of the school board members. As Kermit fire chief, it was alleged, he also ran a "fire for hire" arson ring that gave Kermit the highest arson rate in the state. For those who had fire insurance the money was good.

The Feds would eventually take an interest in all this. In 1986, the FBI organized an undercover operation. The federal lawmen got their hands on a green boxcar, rolled it along the tracks until it had a direct line of sight to the Preece trailer, and sneaked agents in to take pictures through a telephoto lens. They camped out there for two days, photographing more than six hundred people going into the trailer that weekend. Wig figured out pretty quickly somebody was watching and sent his sons and Debbie to surprise the uninvited guest. But when they got there, the boxcar was empty, save for some cigarette butts, soda pop cans, and sweat-soaked undershirts. Now, worry set in.

About a year later, after the bridge leading into Kentucky was blocked off and telephone lines into Kermit disconnected, a convoy of state police cars, backed up by a helicopter, swooped into town. They arrested Debbie and Bull and Wig and Debbie's mother,

Cooney—practically the entire Preece family—that day. They raided Wig and Cooney's home, finding $54,010 under a bed. The federal agents were surprised the Preeces didn't have a bigger stash. Wig's safe was empty, his prized coin collection missing. He never trusted the local bank and its safe-deposit boxes with his money and valuables. He was known to bury things, but he wasn't talking.

All the suspects rounded up would later plead guilty, but Debbie and her husband changed their minds, fired their lawyers, rejected a plea bargain that offered probation, and put their fate before a jury. The trial drew national media attention, and attorney Jim Cagle found himself on center stage. A Georgia native with wavy salt-and-pepper hair and a baritone voice that comforted clients and soothed judges and jurors, Cagle had recently moved to West Virginia's capital, Charleston, after a stint as a lawyer in a legal aid office not far from Kermit. He was dogged when he got onto something, shunning plea bargains and never backing down. He specialized in criminal defense. Debbie told Cagle she was innocent, and he was going to work nights and weekends to keep her out of prison.

For a week at the federal courthouse in Huntington, West Virginia, the government paraded witness after witness—drug runners turned informants, felons seeking to reduce their prison terms, undercover officers—who testified against Debbie and her police-chief husband. The prosecutors—led by a thirty-two-year-old, Harvard-educated assistant US attorney who'd left Boston and followed his wife to West Virginia for her residency in rural medicine—kept asking about marijuana that Debbie had hidden for her mother outside in the trash after Debbie had let that slip on the witness stand. A few times, garbage collectors had hauled the bags to the dump, and digging them out was a messy undertaking. In closing arguments, Cagle told the jury of six men and six women that Debbie had played no role in the family drug business, and to discount the witnesses who'd ratted on her. They had an ax to grind. They had something to get from the government.

"What's your name and what are you in for?" Cagle said in his closing statement. "I've never seen such a lineup of witnesses."

The jury found Debbie and her husband guilty of all charges—drug conspiracy, drug possession, and tax evasion—enough to lock them both up for a long time. Before her sentencing hearing, Debbie tried to orchestrate a deal to avoid prison. She offered to provide free ambulance service in Mingo County for an entire year in exchange for parole. But she needed the cooperation of the Mingo County Commission, and those boys had their own hands full with corruption charges, so they backed out.

A judge sentenced Debbie to ten years in the federal penitentiary, while her husband got fifteen years. The judge also fined her $105,000. Debbie was livid. In her mind, she didn't get a fair trial. Not even close. Outside the courthouse, she blasted the judge, the prosecutors, Mingo County politicians, the whole rotten, stinking, corrupt system. Bribes, kickbacks, nepotism, vote buying, patronage, all of that. It permeated Mingo County, the whole state, and she and her family had been singled out, punished for their politics, treated like dirt.

"If I sold drugs, at least I gave something for their money," she told the media.

The day she left home for federal prison in Lexington, Kentucky, to join Bull and her mother, TV and newspaper reporters waited outside. Her anger hadn't subsided. She swore she would never return to the town the Preece family had ruled like a fiefdom. She would be the last of the Preeces to head to federal prison.

"I'll never live in Kermit or Mingo County again," she said.

The scandal put the Preece clan and the town of Kermit in the pages of *Reader's Digest*, the *Los Angeles Times*, and *People* magazine, which played off the state's unofficial motto with the headline "Almost Heaven? This Corrupt Corner of West Virginia Was More Like the Other Place." National TV reporters also descended on the town to cover the bust. "In the foothills of West Virginia, in Mingo County, lies the grimy coal town of Kermit," reported Mark Potter for ABC News. "Until recently, it may have been as corrupt as any town could be."

For their part, federal officials celebrated the end of the Preece drug ring. For eight to ten years, something terrible had happened in Kermit, the lead prosecutor said at a press conference. Firefighters setting fires, cops selling drugs, politicians buying votes and silence. And now it was over. The Feds had sent a message. More than seventy people snared in the narcotics ring went to jail, nine from the Preece family alone, not counting in-laws.

"This case is the best thing that could happen to West Virginia," said the prosecutor, who would return to Boston. "We reclaimed part of America. The area will never be as bad again."

Debbie didn't blame Cagle for her guilty verdict back in 1987. And besides, prison wasn't so bad after the warden took a shine to her. She was released after three and a half years on account of good behavior. Debbie returned to Kermit, taking a job as a dispatcher for the family trucking company, then ran a motel, restaurant, and bar in East Kermit. One of the motel's guests was a roofer in town helping to build the new K–8 school. They got married. Debbie sold the motel in 2001. She stayed out of trouble.

During her trial, Debbie had grown fond of Cagle—she always called him Mr. Cagle—and she didn't know any other lawyers outside Mingo County, and she didn't trust any of them. Corruption ran deep. So it wasn't surprising that she would call Cagle some twenty years later to talk about her brother's death. She wanted to hold those responsible to account. Sue the bastards. She set up a meeting with Cagle in Charleston. "We'll need evidence," he told her.

In the summer of 2007, Debbie filed a wrongful death lawsuit against Dr. Donald Kiser and Sav-Rite Pharmacy. The complaint, written by Cagle, was four pages. It alleged Kiser knew Bull was an addict but never examined him and wrote him prescriptions for drugs Kiser knew would be abused. The allegations against Sav-Rite were much the same. The pharmacy filled prescriptions that caused Bull's death. It didn't take much convincing for Cagle to take Debbie's case. She gave him a tour of pain clinics and pharmacies in Mingo County. They also learned more details of Kiser's arrest before taking his practice to Ohio. Not only had he been charged with trading prescriptions for sex, but he was also alleged to have written hundreds of illegal prescriptions at the Williamson clinic. Corporal Jim Smith of the Mingo County Sheriff's Office was the arresting officer. Smith's wife worked at Sav-Rite.

At first, Sav-Rite's lawyers tried to get Debbie's lawsuit dismissed. They hired a doctor out of Beavercreek, Ohio, who found no reason to question Bull's prescriptions, vindicating Wooley, Sav-Rite's owner, who filled the scripts. The doctor shifted blame to the addict, writing, "I see nothing wrong with the action of Mr. Wooley filling the oxycodone prescriptions for Mr. Preece. Neither the physician nor the pharmacy has control over how much the patient takes once they have been given a prescription and it has been filled."

The pharmacy's lawyers also raised questions about an August 8, 2007, letter Kiser sent to Debbie from jail while Kiser (inmate #1135259) was awaiting trial. The lawyers had requested copies of any correspondence between Debbie and the doctor, and she was obliged to hand over the letter under civil lawsuit disclosure rules. In two pages, Kiser referenced the "many favors" he had done for Preece, her family, and friends. If by "favors" he meant the illegal prescriptions he wrote for Bull and her sister, then he could rot in jail as far as she was concerned. He asked her to pressure a Mingo couple whom they both knew to pay back $5,000 Kiser had loaned them. "It is time you help me out in my time of need," Kiser wrote from his cell. Kiser also revealed he was going to take a plea bargain to "protect" others. "You know how deceitful and crooked the feds can be," he wrote. "They also threw up Bull's death in my face!"

Sav-Rite's lawyers offered Debbie $15,000 to drop the lawsuit. She wouldn't settle.

4

The Easter Bunny

One Friday morning in June 2008, a private investigator parked his car in the post office lot in Kermit, anchored a video camera on the dash, aimed the lens at the front entrance of Sav-Rite Pharmacy, and pressed record. A light breeze rustled a paper sign that advertised a weekend billiards tournament at the Teardrop Inn. The sun baked in the blue sky. It was a beautiful day to pick up a prescription.

Jim Cagle and Debbie had hired the PI to chronicle the shopping habits of a clientele that hailed not only from West Virginia and Kentucky, but also Ohio, Virginia, and Florida—as license plates attested. The camera captured video of customers—dozens of them—going in and out. Cagle and Debbie were gathering hard evidence for the lawsuit that assigned blame for Bull's death. They figured Wooley would deny any wrongdoing, but the footage would provide proof that he was running a pill mill. They planned to show it to the pharmacist before the trial, surprise him with it while he gave his deposition, see how he would react.

"It was my observation that almost every person coming out of the store was carrying what appeared to be a prescription package," the PI would later write in a report. Sav-Rite patrons were spotted walking up to vehicles and exchanging the contents of the white prescription bags for cash. The buyers held wads of money. Bills for pills. Three or four times, a man in his twenties exited the store with a prescription package and motioned for others to follow him to a back alley behind the three-thousand-square-foot concrete-block pharmacy with the blue roof. The PI couldn't videotape the transactions because customers were too close to his vehicle. A large bath towel atop the dashboard concealed the camera to avoid raising suspicions.

The private eye—Michael Rigsby, of M&T Investigations—drove back and forth from the post office to a nearby convenience store, keeping the camera trained on Sav-Rite. He recorded coal miners, in their blue coveralls fringed with reflective orange and silver tape, stopping to pick up prescriptions. Two women smoked cigarettes while standing beside the entrance. The drive-through window was backed up four vehicles deep—including a van with a BABY ON BOARD sticker and a four-wheeler. Three delivery vans also made stops that morning, their drivers unloading cargo stored in large cardboard boxes and plastic bins. The camera zoomed in on one van's license plate. The Ohio tag was registered to a regional drug distributor, Miami-Luken. The driver was making daily deliveries at Sav-Rite. The stock of prescription drugs required frequent replenishing.

At 2:30 p.m., the investigator packed up his surveillance gear and drove around Kermit, searching for other places to park next time. He found the perfect spot across Highway 52, with help from Tommy "Tomahawk" Preece—Kermit's fire chief and town councilman, and Debbie's younger brother. "It is approximately 600 yards from the store, and is elevated above the area, allowing full view of the pharmacy," according to his report.

There, he later delivered the surveillance tape to Cagle. After watching the video at his office, Cagle was about to do something he had never before done during his forty-year career. He had seen enough on the tape. He had heard enough from Wooley. Two people in Kermit had overdosed during the past two weeks alone. It had to stop. So Cagle started writing a letter to the US

Drug Enforcement Administration. He wrote about the vanloads of Mingo County residents that Kiser's brother-in-law shuttled to Marietta. The trucks and cars with out-of-state license plates that backed up at Sav-Rite's drive-through and packed the parking lot as if a big-ticket country-western act had descended on Kermit. Forget about finding a parking spot at the post office on the first or fifteenth of every month.

Cagle wrote about Wooley's other business activities that the surveillance had uncovered. In addition to Sav-Rite, Wooley owned a used-car lot and had recently placed a trailer on it. A sign outside said FAMILY MEDICAL CENTER. He was offering a free place to stay for roving doctors who worked at pain clinics and were willing to tell addicts to fill their prescriptions at Sav-Rite. What's more, Wooley was building a satellite pharmacy—Sav-Rite #2—a couple of miles up the highway in Stonecoal. It would be attached to a new pain clinic—a one-stop shop for addicts seeking pills. It was brazen, and nobody was doing a damn thing about it.

"The decision to communicate is purely mine," Cagle wrote. "I find the situation deplorable and tragic. It's as if a whole region of people is suffering from an epidemic while a few are prospering while they feed the epidemic and misery which it causes." He signed the letter and dated it. He included a copy of the PI's surveillance tape and tossed the package in outgoing mail. He wanted somebody with a badge to investigate. It was addressed to Special Agent Dominic Grant, the DEA's top investigator in West Virginia.

Debbie and Cagle possessed irrefutable video evidence that something sinister was going on at the Sav-Rite Pharmacy in Kermit. All they had to do was figure out the right moment to spring it on Wooley. Three weeks later, they would have their chance.

After finding a parking spot on a street in downtown Williamson, Debbie and Cagle entered a redbrick building and climbed the stairs to a second-floor hotel suite where Wooley had agreed to be deposed, to answer questions before the trial—a process lawyers call discovery. Cagle could ask Wooley whatever he wanted, and Wooley would have to answer. Or the pharmacist could plead the Fifth Amendment and invoke his right against self-incrimination. But that would raise further questions, more suspicion.

The hotel, Strosnider Suites, stood atop Starters Sports Bar & Restaurant, the most popular restaurant in this coal town of three thousand people, and four blocks from the most popular pain clinic in West Virginia. The Wellness Center churned out more than half of all hydrocodone prescriptions statewide. Kiser had practiced at the clinic before he moved to his new sham of a medical office in Ohio.

Into the suite, Cagle toted a Bankers Box stuffed with manuals, receipts, invoices, labels, and one videotape. Debbie followed closely behind. She was nervous and tried to steady her breathing. Wooley and his lawyer were waiting for them. The suite had a kitchenette—small sink, microwave, fridge. Cagle and Debbie took their chairs at a conference table directly across from the pharmacist. Debbie glanced at the floor, pretending not to notice Wooley's dagger of a stare. She had started to have misgivings about suing him. He was a popular fellow in Kermit, always willing to give a helping hand, loan you the money to buy a used car from his lot, and fill your prescription, no questions asked. Her neighbors in town, even some members of her own family, were whispering that she had filed the lawsuit for money, not for justice. The criticism was getting to her. Besides, she had no guarantee she would win. She was suing on behalf of an addict who abused painkillers. Wooley's lawyer would undoubtedly paint Bull as a criminal, and Cagle worried about how a jury would react. Doctors and pharmacists were typically pillars of their communities in small-town America. They didn't force anyone to abuse pills. The addicts were the ones breaking the law. It wasn't easy to sympathize with them. Cagle realized he had his work cut out for him.

In Jim Wooley's mind, he had done nothing wrong. Sav-Rite

was a drugstore, and a drugstore can't dictate who is and isn't a properly licensed physician—and Kiser was properly licensed in Ohio. Sav-Rite wasn't a health care provider. It had nothing to do with the doctor-patient relationship. The pharmacy had adhered to all West Virginia laws. It could not substitute its opinion for that of a physician's. It could not refuse to fill prescriptions from licensed doctors. Now, the Kermit pharmacist was going to set the record straight.

The stenographer swore Wooley in, then Cagle introduced himself. He was running the show. Debbie was copilot.

"We got cookies and muffins," Cagle said to Wooley.

Cagle knew Wooley was diabetic and was offering him a snack, something to keep his blood sugar up. These questions could take some time. Cagle also was trying to put Wooley at ease. Cagle believed the best depositions were conversations, not confrontations, not interrogations. Sometimes he had to be tough, but he got more out of people with honey than vinegar. They were more likely to put down their shields and open up, talk more freely.

Cagle asked Wooley about his background—a softball question to get things going. Forty-nine years as a pharmacist, Wooley said. Licensed in both West Virginia and Kentucky. But Sav-Rite was the only pharmacy he ever worked at. He was proud of his work as a druggist. He had never been reprimanded, never faced any disciplinary action. The state pharmacy-board inspectors had given Sav-Rite glowing reviews. Wooley kept current his continuing education credits, those CEs, which many pharmacists forgot about and got sternly worded letters from the board about. Wooley's credits were always up-to-date. He had the certificates to prove it. His store manager kept them in a file drawer at the pharmacy, if Jim Cagle or anyone else wanted to see them. The last conference he attended was at Caesars Palace in Las Vegas.

"Was it hot?" Cagle asked.

"I remember having a diabetic episode and falling. And I remember how hard the marble floor was. I remember that very clearly."

"You would absolutely remember that."

Cagle switched gears a bit. He asked about Bull and the medications Kiser had prescribed: Valium, oxycodone, blood-pressure medication. "Would it surprise you if it says you shouldn't take oxycodone if you have a heart condition?" Cagle wondered. Yes, that would be surprising, Wooley said. Sav-Rite subscribed to a computer program that flagged contraindications so the pharmacy would not dispense drugs that were harmful for customers to mix. Wooley said it wasn't uncommon for someone to take OxyContin along with a blood-pressure drug.

"What is the name of that program?"

"Renlar." Wooley spelled it. "R-E-N-L-A-R."

"Who makes it?"

"McKesson & Robbins."

"The big drug manufacturer?"

"Yes."

McKesson & Robbins, the name of the company before it became McKesson Corporation, was a drug distributor, not a drugmaker. Many people conflated the two.

Cagle told Wooley he had a few more questions before they took a break. Their parking meters were running short on time, and the Williamson Police Department wasn't shy about giving tickets.

"Are you doing OK?" Cagle asked.

"I'm doing pretty good."

"I don't want you falling down on the floor like at Caesars Palace. I'm dead serious about that."

"I know."

At the break, Cagle followed Debbie down a staircase to a first-floor lobby. She was crying. Cagle asked her what was wrong, figuring Wooley's deposition was bringing back memories: Bull's battle with addiction, the blood-soaked clothing in the trailer at Mud Lick, the loss of her brother.

"I feel sorry for him, Mr. Cagle."

"Bull was a good man," Cagle said, seeking words to comfort. "I feel sorry for Jim Wooley."

"You feel sorry for *him*?" Cagle responded in disbelief. "He's a lying son of a bitch. Screw him."

After the short recess to feed quarters into the parking meters, Cagle had some items to show Wooley. Debbie withdrew them from the Bankers Box, sliding them across the table. Cagle was still confounded by Debbie's sympathy for Wooley.

The first items were labels from Bull's prescription bottles. They had been peeled off, photocopied, and marked as Exhibit 6.

"Take a second to look at it," Cagle instructed.

Wooley held the paper and stared at the three letters that identified the Sav-Rite employee who had filled the fatal prescription: *JPW*. "My initials are on the label."

"I've got a bigger item for you."

Wooley stared at the thick binder that contained the *Pharmacist's Manual*, 2004 edition. Published by the DEA, it spelled out a pharmacist's rights and responsibilities. Cagle asked Wooley to turn to an appendix, which talked about pharmacists using sound professional judgment, and started reading aloud:

"'The law does not require a pharmacist to dispense a prescription of doubtful, questionable or suspicious origin.'"

Wooley nodded. He was aware of that. At Cagle's request he flipped to page 83, listing a series of bullet points to help pharmacists identify prescriptions that served no medical purpose.

"Did Donald Kiser write significantly more prescriptions or in larger quantities compared to other practitioners in Mingo County?" Cagle asked.

"Yes."

"He's in prison for that, right?"

"Yeah, well, I don't know why he's in prison."

"How many prescriptions would you estimate Donald Kiser wrote that you filled in Kermit, West Virginia, at the Sav-Rite Pharmacy?"

"Thousands."

But Wooley kept insisting he didn't know a thing about Kiser's arrest in Mingo County or his West Virginia medical license being revoked months before Bull's death. Wooley would admit only that he knew Kiser had moved to Ohio. And that he was licensed to practice medicine there.

"But you didn't know he was in trouble here in Mingo County?"

"No."

"And you didn't know the husband of one of your employees arrested him, right?"

"Right."

"And she is your valued employee, right?"

"Absolutely."

"Do you think I believe in the Easter Bunny, Mr. Wooley?"

Wooley's lawyer jumped in: "Objection, argumentative."

"It is." Cagle repeated, "Do you?"

Wooley didn't answer.

On the table was a stack of newspaper articles. Cagle handed Wooley a copy of a story about Kiser's arrest. The newspaper was sold at Sav-Rite.

"Do you believe that it was incumbent upon you, the owner of a pharmacy filling prescriptions, to be aware that doctors were accused of violating criminal laws in prescribing . . . controlled substances?"

"Yeah. I should have made an effort to know, yes."

"And you didn't, I take it?"

"No. All of this is really obvious now."

What an idiot, Cagle thought. Wooley had just admitted his negligence. The confession would bolster Debbie's lawsuit. Just wait till Cagle read back those words to a jury with Wooley on the stand.

Cagle wanted to know whether Wooley spoke to other Sav-Rite employees about Kiser.

"I'm having difficulty here." Wooley explained that he wasn't feeling well, wasn't thinking clearly. He complained he was "uncomfortable." He needed a break. "I'm having health issues. I have it every afternoon, late in the afternoon, due to my diabetes. I don't think I can continue at this time is what I'm saying."

The deposition ended abruptly at 4:45 p.m. Cagle hadn't yet asked Wooley about the video, but he would get another chance in two months. It would be in a new location—at Wooley's attorney's office in Huntington, just blocks from the federal courthouse where Debbie had faced trial two decades earlier. Wooley's lawyer planned to take her deposition on the same day, too. She'd be ready. Cagle would coach her beforehand: listen to the questions, don't be evasive.

Before they left Williamson, Cagle had a surprise for Debbie. He opened the hatch of his SUV, reached into a box, and grabbed a hard-plastic bottle of French champagne with a twist-off top. They retraced their steps to the hotel room, chilling the bottle in a bucket of ice until it was just cold enough to enjoy. Cagle invited the hotel staff to join in a toast. Wooley and his lawyer were already gone. Cagle poured everyone a glass and then clinked Debbie's. Cagle was building a good case, but it was early yet. He'd have to squeeze more out of Wooley.

Round two of depositions started the morning of September 3. From the Huntington law firm's eighth-floor conference room, Debbie and Cagle could see the shimmer of the Ohio River and the convergence of three states—West Virginia, Kentucky, and Ohio. But they had little time to soak in the view, what with Jim Wooley and his attorney sitting across the table. Debbie would go first, Wooley second. After being sworn in, she took a deep breath and braced for what Cagle had warned her would be coming: sharp questions about her family's past.

Wooley's lawyer began by probing Bull Preece's background, his upbringing and education. Debbie explained that Bull had struggled in his early high school years in Kermit, failing most of his classes. His parents, Wig and Cooney, sent him away to a Catholic high school in Columbus, Ohio. He lived there with an older sister. A private school gave him more one-on-one attention, more discipline. After several years, he returned to Kermit to work in the mines as a buggyman, operating loading machines.

"Do you know whether or not your brother had a criminal record?" the lawyer asked.

Debbie glared at Wooley and turned back to the lawyer. He was digging up skeletons. No sense trying to hide the truth. "He does have a criminal record."

"Do you know if he ever actually served any time?"

"He did do time. Yes, he did."

"Approximately when was that?"

"Probably thirty years ago."

It had been twenty-one years to be exact. Time tended to stretch when it comes to things best forgotten.

"Any of your other family members involved in that?" the lawyer asked.

He knew, Debbie thought. The lawyer wanted to get her to put it on the record. "I myself was involved. My sister Brenda. My then husband, David Ramey. My mother. My father. My sister Emma. Her husband, Kenny. Brenda's fiancé, Kerry. That's the ones I can remember now."

The lawyer had Debbie list the drugs sold from the trailer two decades earlier: marijuana mostly, some cocaine, definitely not heroin. She talked about her arrest, her not-guilty plea, her three years in federal prison, and her six months at a halfway house.

"Did any other members of your family serve any time?" the lawyer asked.

"Yes, sir, they did. All of them did."

After a break, it was Wooley's turn. Debbie handed Cagle the undercover video. He wasted no time with pleasantries as in his first go-around with Wooley. The private investigator had caught Sav-Rite and Wooley red-handed. And Wooley would have to

34

answer for it. The tape hummed in a portable video player on the conference table. Cagle let the video run for a minute or two before shutting it off. He let the silence that followed make his point, then turned to question Wooley. Why were carloads of people driving great distances to have their prescriptions filled at a pharmacy hardly much bigger than a double-wide? Why were they passing two dozen other drugstores to get there?

Wooley sheepishly admitted he had a large number of satisfied customers from Paintsville, Kentucky, a forty-minute drive from Kermit and ten times the population. He didn't explain what was wrong with the nine pharmacies in Paintsville. Cagle then slid a two-page report across the table and asked the pharmacist to read it aloud. It had been prepared by the private investigator Cagle and Debbie had hired.

"'Numerous strangers, people who are not regular patrons or residents of your community, suddenly show up with prescriptions from the same physician,'" Wooley recited.

"Is there any particular reason that you know from talking to your customers as to why they would come out to Kermit to fill a prescription?" Cagle asked.

"We compete price-wise. That's probably the biggest issue."

"So you think it's lower prices?" Cagle fought to contain a grin.

"Yes."

Business was indeed good at Sav-Rite. Sales had topped $6.5 million in 2006, according to records turned over by Wooley's attorney. That was nearly double the take of independent pharmacies on average across the country, and no telling how much more than similar-size drugstores in rural towns as tiny as Kermit. Profits had increased each year since Wooley took over the pharmacy in 2002. Sav-Rite had to hire five extra employees.

"I've been doing this a long time, and I'm a competitor," Wooley explained. "We give proper service. We give faster service. We have better service."

Cagle was particularly interested in the faster part. Just how fast? "Do you think that you, maybe, at least as of now, fill as much as a prescription a minute?" At that rate, Sav-Rite would be churning out significantly more prescriptions a day than the number of people in Kermit. It was triple the rate of other independent pharmacies. Even large chain drugstores couldn't keep up with such a frenetic pace. And the overwhelming majority of Sav-Rite's prescriptions were for powerful and addictive painkillers.

Wooley never hesitated. "Yes."

"You do, don't you?" Cagle asked, seizing on a startling admission that the pharmacist delivered as a point of pride.

"Yes."

Debbie gazed out the window to the churning river below. She and Cagle had uncovered more than enough to make their case.

5

Raided

Jim Wooley was fuming. By November 2008, regional supplier H. D. Smith Wholesale Drug Company was blocking some of Sav-Rite's orders for hydrocodone and oxycodone. They had been doing business together for a year. Wooley placed orders, and H. D. Smith always delivered. But now someone at the company was questioning the size of his orders. Wooley called to complain.

Distributors such as H. D. Smith were feeling the heat. They realized the DEA meant business. Joe Rannazzisi's unit was launching multiple investigations across the United States. He was serious about enforcing the law that required the companies to report pharmacies that ordered a suspicious number of opioid painkillers. H. D. Smith had sent its compliance chief to a DEA conference in the fall of 2007 to learn about the agency's expectations. Its executives met privately with DEA brass. Discussions were ongoing. In response, the company developed a "know your customer" program to monitor its prescription painkiller sales. H. D. Smith directed its sales force to collect detailed information on every pharmacy it did business with. The three-page form required a pharmacy to declare the types of patients it served, the names of doctors treating them, and what percentage of total sales were for painkillers. Inspections would follow as needed. It all sounded good on paper, but some kinks were still to be worked out. Jim Wooley's Sav-Rite Pharmacy would put it to the test.

H. D. Smith had good reason to block Wooley's orders. Someone at the company's office in Springfield, Illinois, had placed a handwritten note in Sav-Rite's customer file: "332,500 hydrocodone shipped Feb. 2008" and "2000 Census, pop: 209"—Kermit's estimated population that year. What's more, H. D. Smith had notified the DEA about Sav-Rite. The drugstore's hydrocodone orders were getting out of hand, and a single doctor, Katherine Hoover, who practiced out of the clinic in Williamson, was writing one of every four prescriptions filled by Sav-Rite. In seven months, H. D. Smith reported 106 of Wooley's orders as "suspicious" to the DEA.

The drug orders looked highly questionable, criminal even, but H. D. Smith agreed to hear Wooley out. It sent inspectors to Kermit late that fall. They first interviewed the pharmacy technician who had placed a number of recent orders. She divulged that Sav-Rite was filling six hundred to a thousand prescriptions a day, and 90 to 95 percent were for powerful narcotics such as hydrocodone. How was Sav-Rite supposed to keep up the pace, she asked, if its suppliers were going to curtail deliveries? The inspectors wanted to make sure they were hearing correctly. That number of prescriptions suggested about fifty-four thousand pills were going out of Sav-Rite's door every day. Was that a fair estimate? The pharm tech didn't think twice: "That's right."

H. D. Smith's team next interviewed Wooley in his pharmacy's back office. Could he confirm his employee's claims? Wooley told the inspectors he wasn't sure how many scripts the pharmacy was filling each day, nor did he have any idea what percentage was for painkillers. He admitted, however, that he didn't know many of the customers flocking to his store. And, yes, he blurted out, Sav-Rite did service prescriptions for some doctors with disciplinary issues. The chances that drugs were being diverted for abuse was "likely," Wooley acknowledged. He also mentioned that he had been named in a wrongful death lawsuit, but he didn't say who filed it. All of that was reported back to H. D. Smith's headquarters. All of that. Yet H. D. Smith didn't shut off Sav-Rite. It did the opposite. It resumed shipments to Kermit unabated. The company justified its decision in a note placed in Sav-Rite's file: There were "four hospice centers, two medical centers and four hospitals in the neighboring area." That suggested a "legitimate need" for an increase in pain medications for patients who truly needed them.

The undercover federal agents thought they had seen it all, but not this, not what they witnessed firsthand at Sav-Rite Pharmacy. Cagle's letter to the DEA, along with H. D. Smith's reports, had tipped them off.

Employees were literally throwing prescription bags over the counter to keep pace with the throng of customers. The pharmacy served "courtesy snacks"—hot buttered popcorn was a favorite—to make the wait more bearable. The cash register drawer wouldn't close, it was so full. You couldn't find a spot in the parking lot, day or night. Cars six deep were backed up at the drive-through window. People were splitting their prescription pain pills with strangers outside. They could be heard openly saying they traveled long distances to Kermit because nobody asked questions about prescriptions at Sav-Rite. As a courtesy, the pharmacy workers sometimes stuffed painkiller prescription bottles in white paper bags, and diabetes and blood-pressure medications in brown bags, so regular customers, with legitimate prescriptions, wouldn't risk having theirs snatched by thieves. Everyone knew Sav-Rite was a pill mill. Pill parties were thrown in the parking lot.

The undercover agents were "startled" by what they saw at Sav-Rite, according to an affidavit they submitted to a federal judge. *Startled.* That's the word they used to justify simultaneous raids on Sav-Rite and the affiliated "pain management" clinic—formally called Justice Medical Complex—a couple of miles up the two-lane highway in Stonecoal.

On a Wednesday in late March 2009, federal agents stormed into Sav-Rite, seizing computers and cases of files. Days earlier, they filed a search warrant under seal that revealed they had received a tip that Sav-Rite was "handing out drugs like candy." The search warrant fingered Wooley, doctors, nurses, and physicians assistants in the "get-rich-quick" scheme. The doctors were paid without ever treating patients. One of the physicians was spotted driving around town in a new Mercedes registered to the pain clinic owner, a former Sav-Rite employee. (The woman had eventually transferred ownership of the clinic to her eighteen-year-old son, who was a certified electrician but had no experience in health care.) Wooley worked a deal with the clinic to funnel all prescriptions to his pharmacies. Sav-Rite was filling more prescriptions for hydrocodone than any other pharmacy in West Virginia, Kentucky, Virginia, Pennsylvania, and Ohio. Only Jim Wooley knew how much cash the pharmacy was pulling in.

Federal investigators linked one of those hydrocodone prescriptions to the death of a female patient who paid $150 cash to see a nurse practitioner at the Stonecoal clinic. She was prescribed ninety hydrocodone pills and ninety Xanax on her first visit. She had the prescriptions filled at Sav-Rite #2. The search warrant also mentioned a drug overdose death in September 2005—caused by OxyContin, filled by Wooley at Sav-Rite, prescribed by Dr. Donald Kiser, a "physician known to Wooley 'personally and professionally.'" The agents who typed up the warrant twice used a black Magic Marker to redact the name of the person who overdosed and died. But if you looked close, the last name unmistakably began with a *P*, and the circumstances made clear the deceased's name: William Preece.

Sav-Rite closed after the raid—temporarily—and the pharmacy's illegal business moved elsewhere, a drugstore forty-five minutes down the road.

Exactly a month later, a barrage of email alerts—eight in fifteen minutes—flashed at Cardinal Health's regulatory affairs office in Dublin, Ohio. It was close to five o'clock on a Thursday afternoon, the last day of April. Family Discount Pharmacy in Mount Gay, West Virginia, was struggling to keep up with a mad rush of new customers bearing prescriptions for hydrocodone. The pharmacy, which shared a brick building with the town's post office, was bumping up against Cardinal's monthly sales caps for hydrocodone and oxycodone. It was imploring Cardinal, a giant in the drug distribution industry, to raise the limits. "Due to the closing of sav rite pharmacy located in kermit wv, our volume has increased," the emails asserted. The two stores were thirty miles apart.

Like H. D. Smith and its other competitors, Cardinal, the nation's second-largest drug distributor, had started paying attention to the DEA's warnings. Noncompliance was getting costly. In 2008, Cardinal agreed to pay a $34 million fine to the federal government after the DEA moved to revoke the company's licenses to deliver painkillers from its warehouses in four states—Florida, New Jersey, Texas, and Washington. Cardinal faced allegations that it had failed to alert the DEA about massive orders of the painkiller hydrocodone by rogue Internet pharmacies. The company promised to beef up its standard operating procedures, which already required pharmacies to fill out a questionnaire disclosing everything from the name of the pharmacist in charge to its percentage of customers who paid for their prescriptions in cash. The distributor had also placed caps on pharmacies' pain pill purchases—a source of frustration for high-volume pharmacies such as Family Discount in West Virginia.

By the next morning, the Mount Gay pharmacy's urgent appeals to raise those limits were yielding results. A Cardinal sales representative notified higher-ups that the drugstore had already consented to an on-site inspection and turned over the requisite reports, surveys, and questionnaires, yet Cardinal had inexplicably neglected to raise the purchase limits—set at eighty-five thousand hydrocodone tablets monthly. The Mount Gay pharmacist was upset. The Cardinal sales rep wrote to his superiors, "This account has become very frustrated with the repeated suspicious order monitoring and feels he has provided detailed information to justify his orders."

Cardinal's antidiversion squad, which monitored suspicious shipments, had good reason to keep Family Discount's drug orders in check. It knew the pharmacy in rural Logan County was high risk because of its seemingly never-ending pleas to buy more and more pills. In June 2008, Cardinal had accommodated Family Discount's requests to raise thresholds on hydrocodone purchases on four occasions because of growing sales. At the time, the pharmacy attributed the demand for more hydrocodone to a surge in prescriptions "written by dr. k hoover," whose notorious clinic was about thirty miles south of Mount Gay. By September, two Kentucky pharmacies—both Cardinal customers—reported that they wouldn't fill Hoover's prescriptions. One pharmacist described "lines of people standing outside waiting to get into the office." Hoover was writing more than half of all pain pill scripts filled at Family Discount. She prescribed more hydrocodone than any other doctor in the state. Cardinal did a cursory background investigation. In a memo, a New Jersey–based Cardinal inspector noted Florida's medical board had disciplined Hoover (for writing an "inordinate" number of illegal prescriptions at a pain clinic in Key West), but West Virginia's board never had, despite three malpractice lawsuits filed against her. The memo left out that Hoover was writing a disproportionate number of prescriptions filled by Family Discount.

Its pharmacist made yet another plea to increase order limits in mid-October 2009. Longer wait times at competing pharmacies were driving hordes of new customers to Family Discount, he informed Cardinal. The Kroger pharmacy across the street had started to offer flu shots and was adjusting to a new computer system. Hence, long lines. The closest Walmart, meanwhile, was remodeling its pharmacy. Customers were frustrated. The pharmacist wanted a quick resolution: "Please let me know what I need to do to update our threshold to avoid any regulatory holds for the end of this month."

By January 2010, Cardinal reached a decision. It upped Family Discount's hydrocodone limit to 150,000 pills each month—more doses than most retail pharmacies dispense in a year. The tiny pharmacy in a town of seventeen hundred people suddenly became Cardinal's number one customer for prescription painkillers in the entire state. And no company distributed more pain pills in West Virginia than Cardinal Health.

6

Addicts' Rights

Days before Debbie's lawsuit against Sav-Rite was scheduled to go to trial in the summer of 2009, Wooley's lawyer was given a list of potential Mingo County jurors, a list that included several Preeces and an assortment of their longtime allies. The pharmacy's insurance company immediately agreed to settle. Neither side was allowed to discuss the terms, but court documents revealed that Debbie, as administratrix, received $42,000. The rest of the money was split evenly among her brothers and sisters. The settlement papers ensured that the Preece family could not sue Wooley personally.

After signing the agreement, Debbie's doorbell rang relentlessly. The visitors to her vinyl-sided house beside the railroad tracks had heard about Debbie's settlement. They wanted to know if they could sue doctors and pharmacies, too, and would Cagle represent them. Most were active users. A few brought pictures of loved ones who'd overdosed and died.

They came from Crum and Marrowbone and Kermit, from Beauty, Lovely, and Pilgrim on the Kentucky side of the Tug Fork. There was a scoop operator, a coal-truck driver, and an electrician. There was a homemaker, a grandmother, and Debbie's older sister Brenda, who admitted she used drugs recreationally but didn't become addicted until she started seeing Dr.

Katherine Hoover in Williamson. Their path to addiction often started with a car wreck or a mine accident, and a referral by workers' comp to a rehab center or pain clinic, places with names such as the Wellness Center and Mountain Medical and Aquatic Rehabilitation. They complained of bad backs, torn shoulders, and busted knees, and they found doctors and nurses to write prescriptions for hundreds of pain pills and, for good measure, antianxiety medication to help take the edge off. It sure did. And before they knew it, they were hooked. They went back to the same clinics again and again. Sometimes they saw a doctor, other times not. But they always left with a prescription—and the name and address of the "go-to" pharmacy that would fill it. That was guaranteed. It was as if the doctors, clinics, and drugstores existed only to make them addicts and repeat customers. The number of new addicts was growing. The supply of pills seemed endless. When addicts weren't getting the same high as before, some doctor was always willing to give them an extra boost, prescribing pills of higher strength.

Debbie didn't turn anyone away—even after she was diagnosed with skin cancer earlier in the year. It hit her like an anvil, but it was treatable, the doctors in Huntington had caught it early, and she was going to beat it. It might knock her down, but it wouldn't stop her from helping Cagle gather evidence and round up people willing to put their names down as plaintiffs in a possible new lawsuit. They could band together. They'd have more power in numbers.

So she visited with more addicts, taking copious notes, asking and answering questions about their circumstances. She drove to their houses, up hollows, over rusted one-lane bridges, on muddy roads, just about anywhere, to retrieve prescription receipts, medical reports, workers' comp and disability records—anything to make a case. She passed the files to a Charleston paralegal, who boxed them up and dropped them on Cagle's desk.

In 2010, Cagle filed the first of a wave of lawsuits on behalf of those who had found their way to Debbie's house. The complaints targeted four Mingo County doctors and three drugstores— Sav-Rite, Tug Valley Pharmacy in Williamson, and B&K Pharmacy in South Williamson, Kentucky. More addicts showed up at Debbie's doorstep. These lawsuits, like Bull's case, Debbie and Cagle realized, would be a tough sell to a judge and jury, and they'd be lucky to get that far. The friends and neighbors who stopped at Debbie's house also carried secrets. Many had started using drugs before they showed up at the rogue pain clinics. They stole drugs from family members. They took more pills than they were supposed to. They ignored the written instructions on the bottles. Some sold their prescription pills on the street. They crushed up pills and snorted them. They injected them. They traveled from clinic to clinic, visited multiple physicians, paid in cash, gathering piles of prescriptions. The cops called that doctor shopping, and plenty of doctors were for sale those days.

After two years, twenty-nine Mingo County, West Virginia, and Martin County, Kentucky, residents had filed eight separate lawsuits in Mingo Circuit Court. Cagle represented all of them. The backlash was swift, and not just from the defendants, who argued that addicts—they were criminals, felons, street dealers, a pox on the community—had zero right to sue. Some of the local politicians weren't exactly keen on the lawsuits either. Pharmacies paid business and occupational taxes and sponsored Little League teams. Pharmacists donated to the campaigns of politicians. You hardly ever saw local cops investigating pill mills. Business was business, and the pain clinics and pharmacies were making more than they ever imagined. Debbie was no longer welcome to take her prescriptions for lisinopril, Plavix, and Nexium to Sav-Rite, or to any other Mingo County pharmacy. She had to drive across the river to Kentucky. The inconvenience was minor. Nobody was paying her to find addicts willing to sue. Debbie didn't stand to collect a cut of any settlements. She didn't know where any of this would lead. What she did know was these weren't bad people. They were just like Bull, caught up in something she was still trying to understand. They were all worth fighting for. The survivors. They deserved their day in court.

Lawyers for the four doctors and three pharmacies went straight to the West Virginia Supreme Court, asking it to toss out the lawsuits of twenty-nine recovering addicts. The attorneys argued the suits were frivolous. Two of the five justices agreed. They sought to lock the courtroom doors to drug abusers. Criminals, they argued, shouldn't be allowed to use the courts to profit from criminal activity. The addicts admitted they broke the law. They took more drugs than prescribed. They sold prescription painkillers on street corners and in pharmacy parking lots. They were junkies. And now they were suing doctors and pharmacists, shaking them down in the courts for money that would undoubtedly be used to buy more drugs. Even worse, many of these same addicts were pleading the Fifth Amendment, refusing to answer questions about their own misconduct or to name their dealers. "There are not even remotely innocent victims here," commented Justice Allen Loughry.

Perhaps not, but two other justices took a different view. Shouldn't a jury have a say? Why should a doctor or pharmacy be given immunity from lawsuits simply because their former patients and customers weren't following every single law on the books? Didn't the doctors who wrote an excessive number of prescriptions and the pharmacists who filled them no-questions-asked bear responsibility?

During a Supreme Court hearing at West Virginia University's law school in Morgantown, Cagle had argued that doctors and pharmacists and drug companies were the ones profiting from the prescription drug problem in southern West Virginia. At best, they had their heads in the sand. At worst, they were nothing more than drug pushers who had *MD* or *Inc.* after their names.

They were the bad guys. Not the addicts. Everyone seemed to want to blame the addicts. Meanwhile, the lawyers for the doctors and pharmacists argued that medical licensing boards and the criminal justice system existed to punish bad actors. The drug abusers were the ones who were ducking responsibility.

Not until a year after those oral arguments, in the spring of 2015, did the West Virginia Supreme Court vote 3–2 to allow Cagle's lawsuits against Mingo County pharmacies and doctors to move forward. Five years had passed since Cagle filed the first round of suits. The delay was a reminder that civil cases—especially ones as complex as Cagle's—oftentimes took years to resolve.

Cagle's case caught the attention of state lawmakers. They moved swiftly to pass legislation to keep criminals from filing lawsuits. Under what was touted as "civil justice reform," new laws were established barring convicted felons—or anyone who had even attempted to carry out a felony—from collecting monetary damages from a defendant if their injuries happened while they were committing a crime. The new ban wouldn't affect lawsuits such as Cagle's because they were filed before the new laws were enacted. But West Virginia's courthouse doors, the state's civil justice system, would be locked to drug abusers from that day forward.

7

A Step Up

Debbie had the idea to tail the delivery driver that hauled prescription painkillers into her coalfield community along the Tug Fork.

"Mr. Cagle, why can't we do something to the pharmaceutical companies?" she asked during a stop at his law office in late 2009. The question had been nagging at her. Sure, the doctors and the pharmacists were guilty. Their greed triggered all the bogus prescriptions, but the drugs had to come from somewhere. They were pills, tiny, portable, convenient, marketable, easy to abuse. They had the Food and Drug Administration's stamp of approval. People had this notion that they were safe because they were prescribed. They weren't safe. Debbie surmised the companies making the powerful and addictive painkillers had to be getting rich. Were the same firms also making deliveries? She wasn't sure. The pills weren't just showing up out of thin air. That much she knew.

Debbie was suggesting an entirely new plan of attack, one that he had been thinking about too. It was the next step up in the prescription drug supply chain. A giant step up, he would learn. He asked her how they'd go about tracking down the names of the companies that shipped drugs to local pharmacies.

"I can find out," Debbie assured him.

51

Days later, she took her own prescription to a pharmacy in Martin County, Kentucky, a few miles from Kermit, across the river. The pharmacist apologized. The blood-pressure medication Debbie needed was out of stock. Could she come back? Deliveries came on Tuesdays, he told her, and she let him know she'd be there to pick it up. That Tuesday morning, she was sitting in her car outside the pharmacy. A van pulled up. It wasn't FedEx or UPS or anything like the trucks that delivered packages to homes and businesses. It was an old ambulance painted over, Debbie would later recount to Cagle. No lettering or advertising identified which company the van belonged to. The driver stepped out, grabbed a dolly, and rolled a stack of boxes into the store. On his leaving, Debbie eased her car out of the lot and followed the van across the bridge into Kermit, keeping a safe distance behind so as not to arouse suspicion. The van passed Sav-Rite, which had temporarily closed after the raid, then sped south on the two-lane highway to Hurley Drug pharmacy in Williamson and across the state line to B&K Pharmacy in South Williamson, Kentucky. At both stops, the driver unloaded boxes onto the dolly and wheeled them inside. Only this time there was more than one trip. "Dollies and dollies and dollies," Debbie would remember.

From her parked car, she jotted down the license plate number. A police officer she knew—Kermit's lone cop—ran the tags. The van was registered to Cardinal Health.

Debbie had identified one of the companies that delivered truckloads of pills to southern West Virginia. There had to be more.

One morning in mid-January 2011, Cagle sat across a table from Randy Ballengee, pharmacist and owner of Tug Valley Pharmacy, one of three drugstores the twenty-nine Mingo County addicts were suing. The group had assembled at the Mountaineer Hotel in downtown Williamson. The 117-room hotel opened in 1926. John F. Kennedy had stayed there while campaigning for president in southern West Virginia. Other guests over the years included Eleanor Roosevelt, Henry Ford, Loretta Lynn, and Hank Williams. Now it was popular with overnight tourists who raced four-wheelers on the nearby Hatfield-McCoy trails. Cagle and Debbie, however, had come to the hotel's mezzanine-level conference room to question Ballengee, to take his deposition. In 2007, Ballengee had a tiny cinder-block pharmacy built from the ground up in Williamson, just blocks from the Mountaineer, after securing a $200,000 federal loan through the US Small Business Administration.

"Let me just make sure I understand how this works," Cagle said, turning toward Ballengee. "How do you identify suppliers? How do they identify you?"

"Sometimes they'll call. I guess they'll do cold calls or whatever."

Ballengee was an adversary, yes, but he was being sued by Cagle's new clients, so Cagle could ask him anything, and the pharmacist didn't see the point of withholding the names of his suppliers. He figured Cagle was on a fishing expedition. The more questions Cagle asked about who sold Ballengee the painkillers, and the fewer about filling illegal prescriptions, the better. Cagle didn't understand what he was sticking his nose into. These weren't companies to be trifled with. So Ballengee told Cagle he bought prescription drugs from AmerisourceBergen and H. D. Smith. And when one distributor balked, Ballengee said, another was always eager to fill the void, companies such as Miami-Luken, Keysource Medical, Quest, Anda, Auburn Pharmaceutical, Masters Pharmaceutical, and Harvard Drug Group. Some even offered "daily specials" on pills, short-dated medications, new generics, big discounts.

"If you need something, you can call . . . and maybe get it the next day?" Cagle asked.

"Yes."

"Is the delivery that quick?"

"Yes."

Six months later, Cagle tried to ask Wooley about Sav-Rite's suppliers, but the Kermit pharmacist pleaded the Fifth Amendment to every question. Sav-Rite had settled with Debbie, but that didn't shield the pharmacist from the Mingo County residents that Debbie had rounded up to file the new batch of lawsuits against his drugstore and three others. Even after the raid, Wooley found wholesale distributors, big and small, more than willing to replenish Sav-Rite's inventory of painkillers. The DEA stripped Sav-Rite #2's license but inexplicably failed to revoke the license of the original pharmacy in Kermit. Wooley didn't permanently close Sav-Rite until the end of 2011, after federal prosecutors threatened him with fraud charges, and he realized it was time to make a deal.

On February 2, 2012, Wooley stood in US district judge John T. Copenhaver's courtroom and admitted to filling prescriptions that Wooley knew were fraudulent over a five-month period in 2006. Copenhaver wasn't happy with the plea deal struck by prosecutors and Wooley's criminal defense attorney. The agreement bound Copenhaver to sentence Wooley to two years of probation. Wooley wouldn't have to spend a day in jail.

The eighty-six-year-old judge had recently heard other criminal cases—the convictions of a doctor and a pain clinic worker—linked to Wooley's pharmacy. Copenhaver already knew about Sav-Rite's bad reputation. It ranked twenty-second in the nation for sales of hydrocodone, Copenhaver informed those in attendance. Wooley had cashed in on the scheme, profited from people's misery. The proposed punishment didn't fit the crime, in Copenhaver's estimation. Federal prosecutors explained it was much harder to prove a pharmacist broke the law than, say, a doctor or drug abuser. Pharmacists were a step removed from doctors and patients. Making health care fraud charges stick was difficult. But Copenhaver wasn't buying those arguments. "Don't count on the court accepting this plea agreement," the judge said.

Copenhaver directed Wooley to state for the record the circumstances behind the crime he committed. Wooley explained that he knew the Justice Medical Complex in Stonecoal had closed in 2006 but was still seeing patients there illegally. The pain clinic's employees called prescriptions into Wooley's pharmacy, and he was filling them. A doctor who had worked at the clinic testified in another case that he warned Wooley not to fill prescriptions that were unsigned or phoned in.

Months earlier, Wooley had sold Sav-Rite for $1.5 million and transferred the proceeds to his wife. Sav-Rite filed for bankruptcy protection. The Feds froze Wooley's assets. In bankruptcy filings, Wooley disclosed he owed an estimated $600,000 to a dozen creditors, including drug distributor Miami-Luken and a shredding service.

Over and over, Copenhaver asked Wooley to explain how he broke the law, and Wooley admitted to fraud, forgery, misrepresentation, and deception. The judge accepted Wooley's guilty plea but deferred consideration of the deal that restricted his punishment to probation. Seven months later, Wooley made his second and last appearance in Copenhaver's courtroom. The judge rejected the plea deal. Wooley was given the opportunity to withdraw his guilty plea. He chose not to. Copenhaver sentenced the pharmacist to six months in prison. He was ordered to report to the federal penitentiary in Ashland, Kentucky.

On a May morning in 2012, Jim Cagle was driving his Cadillac Escalade when a black Mercedes shot past him, heading southbound on the West Virginia Turnpike, and for a moment Cagle stopped thinking about how to sue the drug distributors. He got only a glimpse, but Cagle instantly recognized the driver. The bald head. The cigar. They were dead giveaways. It couldn't be anyone else. He hadn't seen Rodney "Bulldog" Jackson in years. He wanted to talk to Jackson, offer him a proposition. Cagle punched his accelerator in hot pursuit.

The chase that morning crossed the border into Virginia, hurtled up knobs of mountains, and descended into valley floors. Jackson was oblivious of his pursuer, never glancing at the rearview mirror. The Mercedes was pushing ninety miles per hour, and Cagle was running the hell out of his SUV trying to catch up. Before they knew it, they were in North Carolina. There, the Escalade finally gained ground, closed up behind the Mercedes, then zoomed up right beside it. Cagle was hammering on the horn. The honking startled Jackson. What the hell was that? Who was this lunatic?

At first, he didn't recognize the madman behind the wheel, shouting out the open window. It was Jim Cagle, wasn't it? The "Georgia Peach"—that's what everyone called him in West Virginia. Jackson's face lit up, and the cigar—one of those expensive Cubans, a Montecristo—dropped from his mouth. A frantic exchange of hand gestures led them to the next exit off I-77 and to a service station where they nosed into the parking lot, shutting off their overworked engines. The two old pals were about to embark on a new quest.

They'd first crossed paths in the 1970s as young lawyers in southern West Virginia. Before Cagle was the Georgia Peach, he was Bubba, his nickname while growing up in Albany, Georgia. He graduated from Albany High the first year it was integrated. His senior class included the gorgeous head cheerleader, Paula Hiers, who would go on to television-cooking-show fame as Paula Deen. After high school, Cagle attended military school at North Georgia College, obtained a political science degree from the University of Georgia, went to law school, then served in the army. In the early 1970s, Cagle landed in Logan, West Virginia, smack-dab in the heart of coal country, taking a job with the Appalachian Research and Defense Fund, a legal aid program. He represented welfare recipients fighting for food stamps and disability benefits. He also did criminal defense work, even though it went against legal aid rules. Cagle had a soft spot for underdogs and the falsely accused. He would eventually make his way to Charleston, where he spent a couple of months as a "Lincoln lawyer," though working out of a Continental would have been a luxury; Cagle made do with his Ford Mercury. A decade later, he was leasing office space at a law firm when it fielded a call from Kermit's police chief, David Ramey, who had been arrested in a federal drug sweep. The chief's wife, Debbie, was in trouble, too. The couple wanted to hire an attorney.

Jackson, meanwhile, also worked as a lawyer in southern West Virginia, in nearby Fayette County, and always seemed to have his fingers in a construction project—most profitable, but some not. Cagle represented Jackson during the first of his three divorces. They'd meet for lunch a couple of times a week at a restaurant called the Smokehouse in Logan. They had the same waitress every time. Jackson, like Cagle, eventually moved to Charleston to practice law.

Now, in May 2012, they were standing in the middle of nowhere, at a gas station in North Carolina, embracing, reminiscing, and both on their way to the South Carolina coast, Cagle to swanky Kiawah Island and Jackson to Folly Beach. They had vacation homes there.

"I didn't think I'd ever catch up to you, you were driving so fast," Cagle said. "What have you been up to?"

"I've been spending eight years drinking whiskey and chasing women," said Jackson, who had gotten the Bulldog nickname as much for his appearance as for his persistence. "But I'm thinking of practicing law again."

"That'd be great."

Cagle needed help. He needed a sounding board, someone with whom to brainstorm ways to sue the deep-pocketed drug companies. Cagle told Jackson about Bull Preece's overdose and the wrongful death lawsuit and the settlement with Sav-Rite's insurance company. They talked about Debbie Preece tailing the delivery truck, Randy Ballengee's testimony about prescription drug suppliers, and the addicts who were suing the doctors and pharmacies.

By this time, Debbie's melanoma had spread. Spots were now on her lungs and brain. She was traveling every couple of months to Duke University in North Carolina for treatments. Cagle's loyal scout, his sidekick, had done her part, but she was sick. She had to put her health first. She needed rest. It was time to pass the baton. Mr. Cagle could find another partner to hold the drug merchants to account until she got better.

"I'm going to get in touch with you," Jackson told Cagle. "I wanna come back. I've been contemplating it. I wanna do it."

"I'd love to have you back. I've got too many of these cases to handle. Come on back. I could use help with the cost. These cases are expensive."

Cagle explained that he was thinking of adding the drug companies as defendants to the existing lawsuits in Mingo County— or have addicts and their families file new suits against the distributors. Jackson nibbled on the cigar, spitting aside shards of tobacco. He didn't like that plan. There had to be a better way.

"You know, with these individual cases, they're not going to do any good," Jackson said. "Let me talk to Rudy and Darrell McGraw."

Rudy DiTrapano was one of West Virginia's most successful trial lawyers. McGraw was West Virginia's longtime attorney general. McGraw had hired DiTrapano's firm in 2001 to sue Purdue Pharma, accusing the company of aggressively marketing OxyContin to state residents, many of whom became addicted. Three years later, Purdue reached a settlement with McGraw's office, agreeing to pay $10 million—a pittance compared to the company's annual revenues. Purdue admitted no wrongdoing.

Jackson suggested that this time McGraw could sue the distributors for violating consumer protection laws. Cagle didn't warm to the idea right away. McGraw was known for steering outside legal work to a select cadre of law firms that donated generously to his reelection campaigns. Cagle worried he would have to do most of the work—depositions, research, finding expert witnesses—but have to split any settlement monies with three or four law firms who stood by to cheerlead. After all, Cagle knew these pill mill pharmacies and pain clinics inside and out. He had the running start.

That's where DiTrapano would come in. He would arrange a meeting with McGraw. Keep everything close to the vest. These were out-of-state companies, big businesses, big pockets. McGraw wouldn't hesitate to sue them. He'd joined the lawsuits against the tobacco companies and won. He'd landed the settlement in the Purdue Pharma suit.

The plan to sue the distributors had an unexpected complication. An attorney at DiTrapano's firm, Josh Barrett, was the brother of Cardinal Health CEO George Barrett. The firm didn't want to be associated with a lawsuit against Cardinal, even though Barrett was planning to retire soon anyway, and he would never take part in the litigation. It made more strategic sense to file two separate lawsuits against the distributors—one against AmerisourceBergen and eleven smaller competitors, and a second against Cardinal alone. You could learn more by trying the cases against the two largest distributors separately, and that's exactly what McGraw did. He appointed DiTrapano, Cagle, and Jackson as special assistant attorneys general to manage the case against the distributors. They'd be paid only if the state won or settled the lawsuits. If it lost, they'd get nothing. They were willing to take the risk.

What Cagle liked about Jackson was that the guy knew how to make money. Cagle was a thinker, the guy who would spend hours on a Saturday at the Supreme Court law library, researching decisions for a kernel of case law that would help his arguments. Jackson was the dealmaker. Besides, Cagle recognized Jackson needed a fresh start. Jackson had been sober going on two years after a dark period of alcoholism that started in 2000, the year his only son died in a car accident. Jackson wasn't exactly known as a hard worker, but Cagle considered him brilliant. Jackson had a knack for making a lot of money for little work. Cagle respected that.

They shook hands in that gas station parking lot in North Carolina. They were going to be partners. Cagle & Jackson, Attorneys at Law. Jackson chomped on the cigar. Cagle smiled. The Escalade followed the Mercedes back onto I-77. It didn't take long for Jackson to shatter the speed limit and leave Cagle miles behind.

PART II

8

A Seismic Shift

It was the last Monday in July 2013, and I was scrambling to finish up a story about a Christian evangelical group's push to outlaw abortion in West Virginia. The group had held a press conference at the West Virginia State Capitol earlier in the day. I was assigned to cover it. My office was in the capitol basement, behind a steel door marked PRESSROOM. The place was a mess, with broken desk chairs, missing ceiling tiles, stacks of old government reports, coffee-stained mugs, yellowed newspapers, a squawk box that blasted lawmakers' floor speeches, and two rat-traps. As a statehouse reporter, I chronicled everything from state agency meetings to corruption scandals. I also pulled a night "cops shift" once a month, writing any breaking-news police stories. I had been working at the *Gazette* for fifteen years, the last three at the two-person capitol bureau. There wasn't much I hadn't seen.

My phone rang. It was Rod Jackson. I recognized his name. About a decade earlier, my newspaper colleague Scott Finn and I had investigated Jackson's business partner, Phillip "Pork Chop" Booth. (We had heard multiple stories about how he got the nickname but couldn't confirm any of them.) Their shell company, National Equity, had fleeced the state out of millions of dollars after floods ravaged schools across southern West Virginia. Pork Chop bribed a state Department of Education's deputy superintendent with vacation cruises and cash. In turn, National Equity received a no-bid contract to replace school furniture, and Pork Chop charged the state four times the going rate and even bought new chairs and desks for several schools untouched by flood-waters. Pork Chop pleaded guilty to fraud charges. He died of a heart attack two months before he was scheduled to be sentenced to prison, and National Equity folded. While Rod Jackson was never charged, our story named him as a company officer. Back then, he didn't return my phone calls seeking comment, but now he was calling me out of the blue.

"I've heard that people are talking about Morrisey threatening to kill our lawsuit," Jackson told me over the phone.

Morrisey was Patrick Morrisey, West Virginia's new Republican attorney general, a forty-five-year-old, gray-haired outsider with a gap-toothed smile. The lawsuit was the one filed against the drug distribution companies in June 2012. Morrisey had inherited it. A New Jersey native, he had moved to West Virginia six years earlier, purchasing a house in Harpers Ferry. He filed to run to become West Virginia's chief legal officer just four days after he secured his law license to practice in the state. He was challenging Darrell McGraw, a twenty-year incumbent.

In midsummer that year, Morrisey became the first candidate during the general election cycle to make a pilgrimage to the capitol pressroom. I stood up at my desk and shook his hand. Our bureau chief, Phil Kabler, was covering the attorney general race. Morrisey, who had resigned as a partner from a Washington, DC, law firm to campaign full-time, outlined a three-pronged platform: legislative appropriation of all settlement awards, no self-promotion through taxpayer-funded trinkets or public-service advertisements, and competitive bidding to hire outside legal counsel. Phil told Morrisey that, because he lived in the state's Eastern Panhandle near DC, his biggest obstacle would be trying to convince southern West Virginians that he wasn't a carpetbagger. That raised Morrisey's ire, and he talked about how his values were more in line with southern West Virginians than McGraw's. He told us he could beat McGraw in November. Not a chance, I thought. Not a chance.

I was wrong.

Morrisey pulled off the upset. He defeated McGraw by sixteen thousand votes. He would become the first Republican attorney general in West Virginia since 1933. State GOP leaders hailed Morrisey's victory as the beginning of a seismic shift in West Virginia politics. Two months later, Morrisey would take office as "the people's lawyer," responsible for protecting West Virginia consumers and all the state's citizens.

As attorney general, Morrisey was now in charge of the lawsuits McGraw had filed against Cardinal Health, Amerisource-Bergen, and a dozen other prescription drug distributors. Jackson was calling about the first case Morrisey ever handled in West Virginia. He had never argued in a courtroom before, as far as anyone knew. He last worked as a lobbyist in Washington, and before that as a staff lawyer for a Republican congressional committee. McGraw had hired Jackson and Cagle as outside lawyers to help with the lawsuits against the distributors. Now, sources had told me, Morrisey was threatening to remove them as special assistant attorneys general and replace them with his own hand-picked lawyers. During the campaign, Morrisey had promised to solicit bids for legal help from outside lawyers. The jobs would go to attorneys who offered to do the work for the lowest price. Cagle and Jackson had signed a contract with McGraw, but Morrisey informed them he was under no obligation to honor it. Jackson begged to differ. The attorney general wanted lawyers over whom he could exercise more control. He didn't trust Cagle and Jackson. They were allies of his enemies, the trial lawyers who supported McGraw during the election.

"You do know Morrisey's wife was a lobbyist for Cardinal Health?" I said.

Two sources had tipped me off about that. But I had no evidence that Morrisey was trying to sabotage either lawsuit by dropping them or settling them with terms favorable to the distributors.

"Can I stop by your office?" I asked.

"Sure," Jackson said.

We made plans to meet within the week.

"And one other thing," Jackson said. "There's a lot I won't be able to tell you."

The walls of the West Virginia State Capitol are thick, lined with pink Tennessee marble that stifles cell phone service, but they seldom keep secrets. Lobbyists, aides, agency administrators, lawyers, secretaries, legislators, analysts—they all trade gossip like currency.

So I wasn't surprised when word had quickly spread that I was investigating the attorney general's ties to the pharmaceutical industry. On the same day, a Democratic lawmaker and an aide to Governor Earl Ray Tomblin both urged me to take a look at Morrisey's inaugural fund. I was flummoxed. Governors held inaugural balls in West Virginia, but I had *never* heard of an attorney general—or any other state official—raising money for an inaugural party. The donation list wasn't easy to find. It was listed on the secretary of state's campaign finance disclosure website as "GOP Inauguration Committee," but this was no celebration for Republicans elected to the legislature or just any state office. This inaugural was to honor Patrick Morrisey.

Who paid for the party? The list named names: individuals, trade groups, and companies. I called our longtime county commissioner, a frequent attendee of fund-raisers. He hobnobbed with philanthropists and political movers and shakers. He was steeped in the daily gossip flowing through the county courthouse, city hall, fraternal lodges, and the capitol. I read the list of donors to him: West Virginia Coal Association, West Virginia Oil & Gas Association, Consol Energy, West Virginia Business &

Industry Council, West Virginia Beverage Association, West Virginia Health Care Association, Cardinal Health . . .

"Stop," the commissioner said. "Stop there."

"Cardinal Health?"

McGraw's lawsuit against Cardinal and other distributors alleged the companies helped fuel West Virginia's problem with prescription drug abuse. As the new attorney general, Morrisey was now the lead plaintiff—and Cardinal Health was helping fund his inauguration ceremony.

There was more. Another firm on the list of inauguration donors was Capitol Counsel. I did some digging. It was a Washington lobbying firm. I checked their client list. They had lobbied for Cardinal for years. I pulled their lobbying disclosure forms filed with the US Senate. They listed the lead lobbyist on the Cardinal account: Denise Henry Morrisey, Patrick Morrisey's wife. She had lobbied for Cardinal since 2002.

I combed through Morrisey's campaign finance reports, flagging $4,000 that Cardinal executives gave to Morrisey—$3,000 of which came after McGraw filed suit against the distributor. One Cardinal exec donated money to Morrisey a month after the election was over.

Then there was the company itself. Cardinal, the third-largest drug distributor in the United States, with $100 billion in revenue the previous year, operated twenty-eight warehouses nationwide and ranked fourteenth on the Fortune 500. Like a handful of its competitors, it had a track record of license revocations and fines. Chief among them was the $34 million fine by the Drug Enforcement Administration in 2008 for failing to notice suspicious hydrocodone orders in Florida. Company brass promised it would never happen again.

But that was a hollow pledge. A subsequent settlement with the DEA over charges of excessive painkiller shipments also in Florida resulted in a two-year ban on Cardinal's license to distribute controlled substances in the state. More fines followed, more promises were made to get a handle on the problem. The company appointed a special panel to review the matter. It declared the distributor had robust systems in place to keep prescription drugs out of the wrong hands.

I sought comment from a Cardinal spokeswoman, asking why the company was giving money to Morrisey's inaugural while fighting a lawsuit overseen by the attorney general. She never responded. I also contacted Capitol Counsel. Denise Henry Morrisey was in a meeting, her secretary said. Follow-up phone calls and an email went unanswered. Her husband wanted his questions in writing, so I sent him an email with questions about the inaugural fund, his wife's lobbying work, and Cardinal Health. An email from Morrisey's personal account popped up in my inbox twenty minutes later. The message was as revealing as it was jarring. Two weeks earlier, I was in Morrisey's good graces. He had sent me an email labeled "scoop," giving me an exclusive on an investigation of a sports media contract at West Virginia University. But this email was not to provide information for my article; rather, it was labeled "not for publication" and took on a decidedly different tone.

> I hereby request a meeting with your editors and publisher to discuss the actual malice coming from your newspaper. I have documented countless errors of intentional bias and would like to sit down to discuss with all relevant folks. With respect to your specific questions here, I find it reprehensible that your newspaper is bringing my wife into your attacks. My wife has nothing to do with any public matters in West Virginia. I am traveling and in meetings today so I would assume that you will not be running anything in the immediate future until we have discussed with the publisher. Thank you.

I forwarded his message to my editors. Did Morrisey intend to file a lawsuit against the newspaper, even before the story was published? He was alleging "actual malice," a legal requirement imposed on public officials who sue newspapers and other media for libel and defamation. To prevail, someone such as Morrisey would have to prove that a reporter and newspaper knew information was false but printed it anyway. It was the first we had heard about "countless errors of intentional bias," the first mention that Morrisey was documenting them. Our editors and publisher offered to meet with Morrisey the following afternoon, but he said he was unavailable and notified us that he planned to bring a lawyer to the eventual meeting. The editors asked him to pick a date and time. "I will get back to you tomorrow with some dates," he responded in an email.

He never got back about the meeting. Two days later, his spokeswoman marched into the pressroom and handed me a typewritten statement. The three pages were undated, on plain paper, not on the attorney general's letterhead. Morrisey alleged that McGraw had talked to him about the Cardinal Health lawsuit at a parade during the 2012 campaign. "My predecessor implied to me at a campaign stop that he had brought suit against Cardinal Health in retaliation for the fact I was running against him," the statement said. McGraw's comments disturbed him greatly, Morrisey said, leading him to believe "at least some part of that case was politically motivated."

I called McGraw. He insisted the conversation never happened, not at a parade, not anywhere. He said he never spoke to Morrisey during the campaign.

Morrisey's statement went on to say that he had stepped aside from the Cardinal lawsuit after the inaugural. "After I took office I decided that, notwithstanding McGraw's comments, West Virginians deserved the case to be decided on the merits," and "while not required under the law, because of McGraw's ethically problematic comments, earlier this year I recused myself from the litigation as it pertains to Cardinal Health."

Morrisey railed against my questions instead of answering them. He accused me of "going into the gutter by bringing my wife into your attacks." He alleged "blatant *Gazette* bias." Political contributions were irrelevant to court proceedings, he said. Again more language that threatened a defamation suit: "When you know something to be false and still print it as an allegation and then follow it with a denial, it represents bad faith." All this, and we hadn't even published the story. In my more than two decades as a reporter, I had never before seen a response like this.

Morrisey also fumed over a question about using leftover inaugural funds to pay $5,000 to GOP consultant Allison Meyers. Meyers had made headlines in 2008 after taking members of the Republican National Committee's Young Eagles, an under-forty-five donor group, to a lesbian-bondage-themed strip club in Los Angeles. Meyers submitted a $2,000 tab to the RNC for reimbursement. She was fired. Now she had surfaced as a consultant to Morrisey. I asked him about her work for the inaugural party. "This is trash," he said in his statement.

The story ran on the front page the next day, revealing the state's largest prescription drug distributor, Cardinal Health, had helped pay for Morrisey's inauguration while Morrisey's office was overseeing a lawsuit that accused the same company of causing West Virginia's prescription drug problem. The article also disclosed that Morrisey's wife was Cardinal's top lobbyist on Capitol Hill.

The following afternoon, I was at the downtown Charleston offices of Robinson & McElwee, one of the most prestigious law firms in West Virginia, where Rod Jackson was leasing a corner office in the law library. Bound copies of state code lined bookcases from floor to ceiling. I took a seat beside Jackson's desk and tried to figure out what exactly he was doing with his ever-present cigar. He wasn't smoking it. There was no smoke or aroma, not even an ashtray. He was practically eating it, chewing the tip and spitting tobacco shards into a paper cup. I did my best to pretend not to notice.

He put his feet up on the desk. They were inches from my face. He wasn't wearing socks.

"People are saying Morrisey's involved in the Cardinal lawsuit," Jackson grumbled.

"Is it true?"

Jackson scrolled through email messages on his iPad until he landed on the one he was searching for. I tried to take a peek. His feet were blocking my view.

"Attorney-client privilege," he said.

Jackson wasn't going to share the email, I realized, but there was nothing to stop me from asking Morrisey for it. Jackson peered over his reading glasses, grimaced, and spat a wad of tobacco. He must have known what I was thinking. I asked, "Did you know he was a lobbyist for two trade groups that represented Cardinal and other drug distributors before taking office as attorney general?"

Jackson nodded.

I needed proof that Morrisey hadn't stepped aside from the state's lawsuit, not just rumors. I sent a formal letter, citing the state Freedom of Information Act, to Morrisey's office, requesting documents, a letter, memos, anything that would show he had recused himself. I wanted to see whether he had separated from the lawsuit, how he had notified his aides. Law firms that face similar situations set up a Chinese wall to insulate lawyers from ethical conflicts. I wanted to see Morrisey's wall.

A thin envelope arrived from the attorney general's office a week later. There was a cover letter, but no documents. "Please be advised that any communication related to your request was communicated orally." Nothing was in writing, nothing memorialized his decision to step aside, nothing was dated. I checked with local lawyers, with officials at the state bar, at least a dozen expert practitioners—none of whom had ever heard of an "oral" recusal.

9

The Pekingese

A week later, I waited for the whistleblower on a sidewalk outside
the Governor's Mansion. Early in the evening now, state employees were long gone, the capitol's gold-leaf dome dimming with
the sun. A red Lexus sedan pulled up to the curb. The driver's-side
window was down. A woman held the steering wheel. A dog, flat-
faced and panting, poked his head from the car.

"You remember Bernard the Pekingese," the woman said.

It was our second clandestine meeting that summer. She was
a former consumer advocate for the attorney general's office.
The newly elected Republican attorney general had fired her
and others after ousting McGraw. Morrisey was now holding
weekly press conferences, promising to burn McGraw's trinkets
in a bonfire and purging office employees—presumed McGraw
loyalists—assigned to the consumer protection division.

I held out my hand for Bernard to sniff. A month earlier, his
owner had leaked to me copies of emails about Morrisey's dis-
mantling of the consumer protection division. On this occasion,
she gave me a DVD. Back at the capitol pressroom, I inserted it
into my laptop, grabbed a bag of pretzels, leaned back in my desk
chair, and watched the footage.

Morrisey was wrapping up his fourteenth stop on a statewide
"listening tour" with a town hall–style meeting in Boone County,

where his predecessor had filed suit against Cardinal and other drug distributors for strategic reasons: Boone had the third-highest prescription drug overdose death rate of any county in the nation. Here, you would find a judge and a jury that had watched opioids lay waste to a community and set off an avalanche of overdoses. The town hall came five days after the *Gazette* published my story. That night, however, no mention was made of Cardinal Health or the state lawsuit that sought to punish drug distributors.

About a dozen Boone County residents sat in the County Commission chamber, a cramped room in the courthouse annex building. A state legislator who represented the county in the House of Delegates stood at the front with Morrisey, who opened the floor to questions.

A Boone County commissioner spoke up. "How do you plan on dealing with the drug problem we have? Right now, people go for employment, they can't pass a drug test. What can we do?"

Morrisey described the problem as "complex." He said he had assigned five office employees to tackle prescription drug abuse. "You move to the supply side," Morrisey said. "You have doctors, nurses, patients, pharmacies, wholesalers, manufacturers. We could go on and on and list these alphabet-soup names. We're trying not to have a game of Whac-A-Mole where we just focus on one opioid and knock it out and everyone goes on heroin. We have to find ways to limit the oversupply. That's a big issue, and that's why I think my background as a health care lawyer would be beneficial."

Morrisey talked at length about his efforts to change West Virginia's reputation as a "judicial hellhole," where juries often sided with fellow West Virginians over large, out-of-state corporations. He mentioned he had had "conversations" with "different companies." He didn't name them. "They just know there's someone they can talk to. At least they'll know they'll be treated fairly now. It's a major step that helps enhance the reputation of the state's legal climate."

After watching the town hall meeting, I was left wondering about the attorney general's conversations. Which companies were talking to him? What did they want? The answer would come on the editorial page of our sister paper, the *Charleston Daily Mail*, owned by the same family, with a separate group of reporters and editors, who worked in a newsroom across the hall from the *Gazette*'s. The Justice Department had blocked the family from combining the two papers, saying the merger would stifle competition. The owners would somehow have to sustain two newspapers until the government's directive expired in 2015. Morrisey had written an op-ed column for the *Daily Mail*, its editorial page a catcher's mitt for conservative columnists. The commentary was about substance abuse. Three-quarters into the column, Morrisey disclosed he had attended meetings to discuss prescription drug strategies with doctors, employers, and "representatives of both Cardinal Health and McKesson, two of the largest drug wholesalers operating in our state."

Morrisey had insisted he had stepped aside from the state's lawsuit against Cardinal, yet he was meeting with the company's representatives. I sent a public records request to his office before I finished reading the column. I wanted emails, letters, memos, anything about that meeting. In response, Morrisey's office released a stack of documents. Among them was a letter sent to Morrisey by Cardinal's lawyer in Charleston. The lawyer had also headed Morrisey's campaign transition team after Morrisey was elected. Cardinal hired the lawyer after firing another—a former US senator, Carte Goodwin, who had handled the case since it was first filed. The letter revealed that Morrisey had met with Cardinal's new lawyer and two company executives in May 2013, five months after Morrisey took office. Cardinal had the letter hand-delivered to Morrisey. It started "Dear General Morrisey" and referenced a PowerPoint presentation describing what the company was doing to stop its opioids from being diverted for illegal use, as well as testimony from the CEO of the trade group that represented drug distributors, the association Morrisey used to lobby for in Washington. It ended with a request for a follow-up meeting with Morrisey, as he had asked the company for additional "materials."

Morrisey declined to answer my questions about his meeting with Cardinal executives and their lawyer. Instead, his spokeswoman issued a statement, defending her boss, reiterating he had stepped aside from the Cardinal lawsuit "when he was not required to do so."

I reread Morrisey's column, which seemed designed to paint his meeting with Cardinal as a noble endeavor, a chance to work together with drug distributors to curb the proliferation of prescription drugs, a first step toward solving the opioid epidemic. West Virginia's largest shipper of painkillers only wanted a seat at the table. But characterizing the meeting that way was problematic. The letter's header disclosed the reason for Morrisey's May 2013 meeting with Cardinal Health representatives. There, at the top, were the words that made the meeting's purpose indisputable—a reference to a lawsuit filed in Boone County Circuit Court. It cited the case number: 12-C-140, the state lawsuit against the distributors that McGraw had initiated but was now in the hands of a new attorney general with seemingly different motivations. The header read, "Re: State of West Virginia, Patrick Morrisey v. Cardinal Health."

The Boone County Courthouse was closing in five minutes. I hurried across State Street, sidestepped the bronze statue of a coal miner on the courthouse lawn, squeezed through the metal detector, and handed the case number scribbled on notebook paper to the clerk behind the glass window: 12-C-140. She fetched the file and made a copy. I paid $33 in cash.

I didn't wait to return to Charleston, a forty-minute drive on the four-lane highway known as Corridor G. I sat in my car in the gravel parking lot between two mud-spattered pickups and started to page through the lawsuit. There were general accusations. Cardinal delivered prescription drugs. West Virginia was the nation's "most medicated state." Pharmacies filled more prescriptions per capita here than anywhere else. Overdose death rates had quadrupled. The drug crisis was costing the state $430 million a year—babies born addicted, families destroyed, skyrocketing jail bills, hospital emergency rooms overrun with patients seeking medication for pain. Cardinal's run-ins with the DEA in Florida were listed. The company's conduct in West Virginia was no better, the lawsuit alleged. Cardinal supplied painkillers to rogue drugstores that filled bogus prescriptions written by unethical physicians. The complaint singled out one such pharmacy, but provided no evidence that Cardinal shipped drugs there. The pharmacy was making $500,000 a month in sales. Nearly 90 percent of the drugs ordered were prescription painkillers such as oxycodone and hydrocodone. It paid drug shippers hundreds of thousands of dollars per year.

I turned the page. A line on page 3 made me pause. I checked it twice, making sure I read it correctly. It was a number: 3,194,400. The pharmacy received 3,194,400 tablets of hydrocodone in 2006. Millions of pain pills—Lortab, Vicodin, Percocet—sold to a single pharmacy in one year. The pharmacy had filled prescriptions at a rate of one per minute. The lawsuit didn't name the pharmacy. It only identified the location: Kermit, West Virginia.

10

Sustained Outrage

Elizabeth "Betty" Chilton was fighting to keep her newspaper from going under. By 2013, the *Charleston Gazette* was in financial distress. It was missing loan payments. Legal bills were stacking up. Advertising revenue was plunging. The number of subscribers was dropping by two thousand to three thousand a year. The Internet was killing newspapers across America, and the *Gazette* was no exception. Google and Facebook were gobbling up retail advertising dollars, while eBay and Craigslist were doing the same with classified ad revenue. Newspapers were slow to react to the assault.

To make matters worse, the *Gazette* had a $30 million liability. The debt financed the 2004 purchase of its rival, the *Charleston Daily Mail*, for $55 million, a deal that could go down as one of the worst business decisions in West Virginia history. The US Department of Justice, meanwhile, had its foot on the *Gazette*'s throat, forcing it to keep the *Daily Mail* in business as a separate newspaper. Two newspapers owned by the same family were competing against each other. Chilton and the *Gazette*'s executive editor, Rob Byers, tried to shield reporters from the newspaper's financial troubles, but it's hard to keep secrets from reporters, and signs that the newspaper was struggling to stay afloat were everywhere. Nobody was getting a raise. The Christmas bonus had gone from the equivalent of two weeks' salary to a $100 bill and, later, to cheap pizza and a Kroger vegetable tray. Soon, we also discovered that the pension plan was in jeopardy. Byers had to justify filling a reporter's position every time one came open. Reporters had to ask special permission to travel to outlying counties to cover stories.

For more than forty years, the *Gazette* and the *Daily Mail* had operated under a "joint operating agreement," sharing advertising and circulation departments but keeping their newsrooms independent. The odd arrangement had benefited readers. The competition spurred more stories, more coverage, and sustained two distinct editorial pages—the *Gazette* being the liberal voice, the *Daily Mail* staunchly conservative. The *Gazette* was distributed in the morning, the *Daily Mail* in the afternoon.

So when the *Daily Mail* went up for sale in 2004, Chilton, the *Gazette*'s seventy-six-year-old publisher and one of Charleston's most generous philanthropists, had a decision to make. She could let the *Daily Mail* be sold to the Ogden Newspapers chain, a longtime competitor that owned daily newspapers across West Virginia and wanted a presence in the state capital, Charleston, or she could buy it herself. Under the joint operating agreement, the *Gazette* had the right of first refusal to buy the *Daily Mail*, and Chilton exercised that right. The $55 million price tag seemed outrageous. The *Daily Mail*'s circulation had dipped to thirty-three thousand copies, and the going rate for newspapers at the time was about $1 million for every thousand subscribers. Still, Chilton wanted to keep the paper out of the hands of a competitor—and rumors were flying that a Charleston TV mogul was also interested in buying the *Daily Mail* if Ogden backed out. The money-losing *Daily Mail* suddenly became a hot commodity.

Chilton privately told me her decision ultimately came down to peace of mind. She could rest easier knowing the family controlled both papers. She figured there would be fewer headaches, fewer distractions. No more disputes with competing owners who ran the newsroom across the hall. With one owner, the newspapers would have the best chance to succeed. Or so she thought.

The first time I stepped into the *Gazette* newsroom, in 1998, two words came to mind: *fire hazard*. There were no windows. Old newspapers, books, and boxes were strewn everywhere. The coffee machine was smoking. The cubicles were falling apart. The desk chairs were broken, wobbly, and stained. It didn't make the best of impressions when I interviewed for a reporter's job.

Amid the clutter, I spotted a copy of an article, tacked to a corkboard, with a headline that grabbed my attention: "W. E. 'Ned' Chilton: The Hallmark of Crusading Journalism Is Sustained Outrage." The November 18, 1983, commentary, published in the *Gazette*, chronicled a speech Chilton had given earlier in the year at the Southern Newspaper Publishers Association convention in Memphis. The *Gazette*'s fiery publisher told his peers that most papers display mere "spurts of indignation. . . . We hit an issue and then pass on to something else. We show the attention span of a postal clerk." Chilton implored newspapers to vent "sustained outrage over basic injustices." He also directed criticism at editorial pages. "Our editorials make the sound of a decorous jackhammer, not the startling thump of a sledgehammer, and, worst of all, we don't keep hammering away, day after day, day after day."

Ned Chilton, whose family had owned the *Gazette* since 1902, pounded away with his sledgehammer press during his two decades as publisher. His piercing blue eyes put a scare in reporters and public officials. He was always fighting for the underdog, denouncing injustices, and pushing for changes that affected everyday people. He hated corruption. A favorite target was Governor Arch Moore, whom Chilton accused of being a crook and unfit for office. Moore, in turn, dubbed Chilton's newspaper the

"Morning Sick Call" because of its relentless "negative" coverage. Chilton filed lawsuits against government boards to stop them from meeting in secret. He sued the state bar and the Board of Medicine to reveal complaints against lawyers and doctors. He advocated for tougher laws against drunk drivers. He attacked cigarette makers, warning readers about the death toll caused by tobacco. He advocated for the integration of schools, pools, restaurants, and clubs. He blasted the state Human Rights Commission when it dragged its feet on complaints. He criticized the FBI's handling of investigations into the deaths of civil rights activists in the Deep South. Such pursuits prompted FBI director J. Edgar Hoover to keep confidential files on Chilton and the *Gazette*, and when Chilton found out, he used the Freedom of Information Act to pry loose the records and publish stories about them. Hoover complained that the *Gazette* had been "consistently hostile" to the FBI.

In 1987, four years after his call for "sustained outrage" at the publishers' conference, Chilton died of a heart attack while competing in a national squash tournament in Washington, DC. He was sixty-five. His wife, Betty, and eighteen-year-old daughter, Susan, were planning to fly there to meet him after the tournament. After the funeral, Betty Chilton fielded phone calls from newspaper chains interested in buying the *Gazette*. It was worth an estimated $90 million to $100 million at the time. The chains were gobbling up family-owned papers across America. The sale would set up the Chilton family for life. And what did Betty Chilton know about running a newspaper? She was a Charleston socialite, more comfortable at the neighborhood garden club than in a smoke-filled, coffee-fueled newsroom. The offers rolled in, but she dismissed every one of them. She decided she was going to learn the business. She was going to honor her husband's legacy.

I made a copy of the 1983 article about Chilton's speech. I read it on the plane ride home to Florida, where I was finishing up graduate school. I had another job offer from a newspaper in Arkansas, but something about this small newspaper with a dozen reporters in the heart of Appalachia was more appealing. There would be night police shifts, dubious assignments, government meetings that stretched for hours, incessant calls for daily stories to fill the news hole. But there would be opportunities, too. The *Gazette* valued investigative reporting. It covered the state without fear or favor. The newsroom was full of journalists who believed in holding the powerful accountable, reporters who persevered in the face of people, obstacles, and forces that threatened to derail their efforts. This hometown newspaper instilled a David-like confidence that when you're on the right track, the Goliaths of money and power could neither intimidate nor stop you. It was keeping sustained outrage alive.

By buying the *Daily Mail*, Betty Chilton thought she could better control her company's destiny, but before she got much of a chance, someone sicced the Justice Department on her, and the government was telling her what she could and could not do with her new acquisition. What Chilton couldn't do was shut down the *Daily Mail*. She could, on the other hand, undo the 2004 purchase, go back to square one, and put it back up for sale—at a huge loss. Or she could prop up the *Daily Mail*, spend millions of dollars to restore it, just so it could compete with the other struggling newspaper she owned.

On May 17, 2007, the United States of America, under the direction of Attorney General Alberto Gonzales, filed a lawsuit against the Daily Gazette Company—which owned the *Gazette* and the *Daily Mail*—and MediaNews Group, the Denver-based company that had sold the *Daily Mail* to the Chilton family three years earlier. The lawsuit alleged the deal, which gave the Daily Gazette Company the unilateral right to shut down the *Daily Mail*, was illegal. The government was demanding that Charleston, population fifty thousand, remain a two-newspaper town.

The lawsuit listed the steps the Daily Gazette Company took to ruin the *Daily Mail*. Its newsroom budget was slashed. The company stopped discounts and promotions for the newspaper, halted delivery to thousands of subscribers, and ceased publication of its Saturday edition. It also pushed to convert *Daily Mail* subscribers to *Gazette* subscribers. Half of the *Daily Mail*'s newsroom staff left the paper, and they weren't replaced. Three of them—an outdoors writer, a photographer, and a college football reporter—came across the hall to work with us at the *Gazette*. The *Daily Mail*'s circulation plummeted from thirty-five thousand copies to twenty-four thousand. It was all part of a master plan—recommended by high-dollar consultants, shared with lenders, and implemented by *Gazette* management—to cripple the *Daily Mail*, to make it a "failing firm," the definition of which under federal law would give Betty Chilton and her family the grounds to close one newspaper, the *Daily Mail*, to save another, the *Gazette*.

After *Gazette* managers got wind of the federal antitrust investigation, they scrambled to stop the damage to the *Daily Mail*, but it was too little, too late. The paper was a shell of its former self. Chilton's lawyers argued that the demise of the *Daily Mail* was inevitable. Only twenty cities in America still supported two newspapers. It was the first time the Justice Department had sued a company under these circumstances. On some days, the front pages of the *Gazette* and the *Daily Mail* were nearly identical. The stories were the same, just written by different reporters. We were covering the same government meetings, the same press conferences, the same fires, murders, and drug busts. The sports pages also frequently mirrored each other. Both papers were now delivered in the morning. "In this lawsuit, the Justice Department declares that it has set out to rescue the *Daily Mail*, but its misguided attempt is more likely to destroy the *Mail* than to save it," Chilton's lawyers wrote in a motion to dismiss the lawsuit.

After three years of legal wrangling, costly depositions, the retention of expert witnesses, and escalating bills from Washington and lawyers hired by Chilton to make the lawsuit go away, the Daily Gazette Company settled with the federal government. Chilton agreed to keep the *Daily Mail* in operation for at least five years. Chilton's company would pay a $225,000 annual "management fee" to MediaNews. The Daily Gazette Company also would have to offer *Daily Mail* subscriptions at half price.

But the settlement didn't turn around the company's finances. It wasn't making its loan payments on time and opted to refinance. Only this time, to secure the debt, the bank placed a lien on the newspaper's 192,000-square-foot building in downtown Charleston, the adjacent five-story parking garage, and its Goss Metroliner printing press. The new loan topped $20.7 million. In the *Gazette* newsroom, we braced for layoffs, for massive budget cuts, for a possible sale. Everyone knew both newspapers couldn't be sustained. Things were going from bad to worse. The *Morning Sick Call* was on life support.

Then something unexpected happened. We heard we would be getting new equipment—video cameras, new lenses, a newsroom content-management software system described as "the Cadillac" of such applications, which would better organize our stories and improve our website. It was even rumored that some of the younger reporters might be getting raises. It was as if the newspaper had found a secret benefactor, a lifeline, a hidden pot of money.

The *Daily Mail* newspaper of London had made the spending spree possible. The *Charleston Daily Mail* had been first to claim the dailymail.com domain name, and now the British newspaper wanted to buy it. Readers hoping to get the latest news from the United Kingdom were being directed to the latest news in West Virginia. So Chilton and the Daily Gazette Company sold the dailymail.com Web address for $1.5 million, one of the twenty-five highest prices ever paid for a domain name. At least $1 million from the sale was used to pay down debt, and the bank refinanced the loan yet again. The rest was invested in the newsroom and used to pay off overdue management fees owed to MediaNews, the old owners of the *Daily Mail*. Both newspapers would survive—for a time.

This startling turn of events was soon followed by a demoralizing one. Chilton and the Daily Gazette Company, court records would later show, didn't own the *Daily Mail* website address. They sold something that never belonged to them. MediaNews claimed ownership. The *Gazette* was not out of the woods, far from it.

11

The Chase

When I stepped out of the Boone County Courthouse and spotted Dan Greear, the attorney general's chief counsel, huddling with a gaggle of lawyers at the bottom of the hill, I flipped open my reporter's notebook, clicked on my pocket-size Sony voice recorder, and hurried down the steps.

It was the middle of October 2013, two months after my story about Cardinal Health's helping to pay for Morrisey's inaugural party, and I was chasing after Greear outside the courthouse in downtown Madison to ask him why his boss was meddling in the lawsuit against Cardinal. Greear was Morrisey's top lieutenant. I wanted to know why Morrisey wasn't telling the truth about his role in the case. I wondered whether Morrisey was negotiating a favorable settlement for the company, or, as his critics believed, trying to get the suit dismissed altogether. Morrisey wasn't talking to me. Maybe Greear would. This was my chance to find out. It's easy to ignore a reporter's phone call. It's much more difficult to duck questions when confronted in person.

Moments earlier, Greear and I had attended a hearing inside the courthouse, listening to lawyers for AmerisourceBergen tell a judge that the state had no right to inspect their pill-shipment invoices. Cagle and Jackson, who were representing the state for Morrisey's office, were there, too. They wanted copies of the sales records, which would presumably show daily and monthly pain-killer deliveries to pharmacies across West Virginia. The invoices would provide Cagle and Jackson with ammunition to make their case that the companies were fueling the state's prescription drug problem, putting profits over the public's health.

Now that the hearing was over, everybody was headed back to Charleston. I caught up with Greear as he prepared to cross the street, waiting for traffic to pass. The other lawyers scurried away. That left the two of us. Greear, who ran for attorney general in 2008, losing to McGraw by just four thousand votes, was forty-four years old but looked much younger. His blond hair didn't have a speck of gray, and he kept fit by riding long distances on a road bicycle.

"Is the AG recused from both cases or one?" I asked, referring to the lawsuit against Cardinal and the separate one against Ameri-sourceBergen and a dozen other prescription drug distributors.

"He is recused from Cardinal," Greear said.

"How long have you been managing that case exclusively?"

"I've been managing it since the first of the year. I always have been managing it."

"So he hasn't had any involvement since day one?"

"Day-to-day? No. I've been doing day-to-day since the first of the year."

"So he hasn't played any role, had any voice in any decisions?"

"I've been doing the day-to-day handling for it all along."

"Well, when was the last time he was involved in the case?"

"I think we're talking past each other."

"What do you mean?"

Greear didn't answer. There was a long pause. I got the sense that Greear was trying to be honest but wasn't going to divulge anything that would put his job in jeopardy.

"Does the AG still have input in the case?" I asked.

"No."

"Well, when did that stop?"

"I have always been managing it from day to day. I'm not

going to get into when he formally recused and when he did all the stuff. I'm telling you, I've always been managing this case since I got into the office."

"But he has had input up to a certain point?"

"Um, I've always been managing, and that's as far as I'm going to go."

"Why won't you answer the question? I mean, it's not—"

"Number one, it's a process. You go through Beth Ryan." She was Morrisey's press officer. "I'm not going to be speaking for the office. That's the process you go through."

Morrisey had enacted a strict media policy. All questions had to go through the communications director. McGraw's office didn't have a communications director—or any spokesman. McGraw's top aides fielded reporters' inquiries.

"When were you told orally when the AG recused himself from the case?"

"I can't pull the date out for you," Greear said.

"January fourteenth, May third, the month?"

"I'm not going to mess up—I mean, I'm not going to pull a date out of the air because I don't know. What I can say is, I've always been managing this case since I got here as part of the things I do."

"But can you say was it last January?"

"No."

"Why not?"

"Again, I'm not the press person."

"OK. But it seems like an easy question?"

"I understand. I understand."

I followed him across State Street, the main thoroughfare through town, to the parking lot at the Madison Volunteer Fire Department. Gravel crunched under our shoes. I kept the conversation going.

"I know what you're asking," Greear said. "I'm going to let the AG parse those things. I know exactly what you're asking. I understand. I'm not going to get my butt in—"

Here, outside the courthouse, Greear wasn't going to tattle on his boss to a pesky reporter, but he wasn't going to tell lies either. The interview ended in the parking lot. But it wouldn't be the last time Greear would be questioned about his boss's role in the Cardinal case.

After Greear drove off, I tapped the red stop button on the recorder to shut it off, fumbled with my car keys, and turned the door lock.

"Are you Eric?"

Someone was standing behind me. I recognized the voice instantly, baritone with a Deep South accent, the voice that had just argued with the corporate lawyers in the courtroom. It belonged to Jim Cagle. Morrisey hadn't fired Cagle and Jackson after all. Morrisey hadn't torn up their contract. They were still working on the lawsuits against the drug distributors for Morrisey's office. They remained special assistant attorneys general. That infuriated Morrisey, but because of my ongoing coverage, the attention was too much, the heat too much, to cut them loose right now. I had put a spotlight on Morrisey and his lobbyist wife and the lawsuits against the distributors through a series of stories on the front page of the state's largest newspaper.

Cagle and I had spoken on the phone a couple of times in recent weeks—he repeatedly declined to comment on the Morrisey coverage, but we had never met until now. I turned around, nearly lost my balance, and shook his hand.

"We need to talk sometime," Cagle said. "How about lunch?"

Misha Tseytlin (pronounced "SATE-len"), the Russian-born, black-bearded general counsel to Attorney General Patrick Morrisey, stood in a Charleston courtroom, fighting to block the *Gazette* from getting its hands on an email about his boss. Tseytlin had previously disclosed that the email mentioned the attorney general and Cardinal Health by name. It was sent five months after Morrisey took office. But that was all Tseytlin would reveal. He

characterized the email as "highly confidential," and he was asking
Kanawha County circuit judge Charles E. King to keep it that way.

It was 11:10 a.m., May 20, 2015, and I was in King's courtroom to report on the hearing. Morrisey had resisted releasing
the email, the same one that the cigar-chomping Rod Jackson
had refused to show me two years earlier, when we first met at
his office in Charleston. Tseytlin, a former clerk to US Supreme
Court justice Anthony Kennedy, spoke from the defense table.
Patrick McGinley and Suzanne Weise—husband and wife, both
law professors at West Virginia University, who were representing the *Gazette*—sat at the table opposite. In the spring of 2014,
our newspaper had filed suit against Morrisey, after he refused
to release the email and other records we suspected would shed
light on his intentions in the Cardinal lawsuit. Did he plan to
strike a backroom deal with Cardinal's lawyers and settle for pen-
nies on the dollar? Was he going to let his wife's client off the
hook? Morrisey alleged our complaint was designed to "impugn"
his character because the *Gazette* had a cozy relationship with
McGraw and couldn't get over that he had been voted out.

The sixty-eight-year-old judge assigned to the case was no fan
of the press. Charles King created a public uproar in 2008 when
he retired and ran for office again, taking advantage of a loophole
in state law that allowed him to collect both his state pension and
judge's salary. He hadn't forgotten the *Gazette*'s coverage of his
double-dipping.

In a court filing, the *Gazette* had asked Judge King to read
the email in private and determine whether it should be released.
Tseytlin was arguing against that.

"What is your name again, sir?" King asked.

"Misha Tseytlin."

"How do you spell that?"

"M-I-S-H-A, last name, T-S-E-Y-T-L-I-N."

"Tseytlin," King repeated, pronouncing it "Satan." "Tell me
something."

Morrisey's thirty-four-year-old lawyer had a lot to tell. He spoke uninterrupted for the next ten minutes, his Russian accent still mildly detectable. He had immigrated to the United States with his family when he was seven years old. He described the email as "sensitive litigation strategy." Other withheld documents, Tseytlin explained, had to do with Morrisey's seeking advice from a retired ethics professor—and those were private conversations, not public records, he asserted. Nothing in the attorney general's files, Tseytlin insisted, would reveal whether Morrisey had stepped aside from the Cardinal lawsuit—or when.

After Tseytlin rested his argument, the judge called on McGinley. A law professor in his late sixties, whose gray hair cascaded down to his shirt collar, McGinley had been arguing public-records cases in West Virginia for decades. He had agreed to represent the *Gazette* free of charge, a move that caused friction with university administrators, who wanted to keep a cordial relationship with the state's chief law officer. Over the years, McGinley had also taken on coal companies that destroyed the land and endangered miners. He argued the first lawsuit that challenged the strip-mining practice called mountaintop removal—and won. He'd represented citizen groups that successfully fought to preserve wilderness areas. And he served on governor-appointed panels that investigated the Sago and Upper Big Branch Mine disasters, concluding that a combination of mine safety failures by coal companies and lax enforcement by regulators caused the explosions killing forty-one miners. During hearings, McGinley asked the coal bosses tough questions, something they weren't accustomed to. "Who does he think he is, Clarence Darrow?" commented one government investigator, referring to the famed early twentieth-century lawyer.

Now, McGinley was on the receiving end of questions from a judge who wanted to know why the newspaper was suing the attorney general, whose duties included enforcing public records laws for all state agencies. Morrisey's office even published a how-to guide on Freedom of Information Act compliance.

"It's been two years—almost two years," McGinley said, referencing the time that had elapsed since I first requested the email that mentioned Cardinal Health. "There has been a shifting target from July 2013."

Morrisey's office had changed its description of the email's contents three times. McGinley told the judge the *Gazette*'s lawsuit was more than a fight for a single email: "This matter began when a *Gazette* reporter made inquiries about what is an enormously important issue here in West Virginia, the proliferation of prescription drugs and the misuse of those drugs. Cardinal Health is one of the major national distributors of prescription drugs."

Since July 2013, I had been asking questions about Morrisey's role in the Cardinal case because he was a lobbyist for the pharmaceutical industry before he became West Virginia's attorney general. And it was odd, McGinley said, that Morrisey told his staff he was stepping aside from the case but never put it in writing.

Judge King wanted to hear more about that from Tseytlin. "Did the attorney general do that?"

"He voluntarily stepped aside from the case to avoid any appearance of impropriety."

"Well, all right. Did he recuse himself or just say, 'I'm not going to have anything to do with it; you all'—meaning the staff—'take care of it'? What?"

"Your Honor, I wasn't there at the time. My understanding is that he said, 'I don't want—I want to be rolled off from this case because I don't want any appearance of impropriety.' I mean—"

"Well, was it because he felt he had a conflict?"

"No, Your Honor. But because of—because, you know, he is a public official and he wanted to be beyond reproach, so he decided he would step away."

The previous year, I almost got a peek at the email that Tseytlin was fighting so hard to keep secret. A Morrisey aide invited me to the office on April 1 to see it, but with much of the text blacked out. I made plans to be there. That morning, the aide emailed me to say that I could only look at the email, not make a copy or snap a picture with a smartphone. McGinley called Morrisey's aide to protest the restrictions. Two hours later, we received notice from Morrisey's office. They canceled my appointment. They promised to get back.

"Well, they never got back to us," McGinley told Judge King. "Waited four months and we filed this lawsuit. This is a newspaper trying to do its job, not, as the attorney general argues, this is some kind of political hatchet job."

King directed Tseytlin to turn over the email and other records to the court. King would read them in his chambers and determine, within the next couple of months, whether the attorney general must release them.

"Do you have them with you here today?"

"I do, Your Honor," Tseytlin said.

"I assume you would want that under seal?" McGinley asked, figuring the records would be secured in the court file, so only King could inspect them.

"I was going to see if Mr. Tseytlin wanted to give you copies," King said.

McGinley picked up on the joke and smiled. "Well, I would be very appreciative."

"Do you want to provide them copies of what you send me?" the judge asked Tseytlin.

"I do not, Your Honor. Those are highly sensitive."

At the hearing's end, Judge King asked Tseytlin and McGinley to write up proposed rulings and submit them to the court. McGinley's proposal would have the judge ordering the release of the email. Tseytlin's would keep the email message forever hidden.

"Any time frame for the proposed orders that Your Honor would like?" Tseytlin asked.

"Don't drag your feet," the judge said.

12

Hunger Games

Morrisey retaliated against the newspaper with a vengeance. He had a subpoena hand-delivered to the newsroom. He demanded we turn over our financial statements and personnel records. If we didn't comply, he was taking us to court.

Two days earlier, on July 19, 2015—exactly five years after it signed the decree with the Justice Department to keep the *Daily Mail* up and running—the Chilton family had announced that it was combining the *Gazette* and the *Daily Mail* to form the *Charleston Gazette-Mail*. Susan Chilton, Ned and Betty Chilton's daughter, was named publisher. The *Gazette-Mail* would feature two distinct editorial pages. Everything else would consolidate.

At 8:44 that evening, Morrisey jumped on Twitter: "Oversight of competition among businesses is a fundamental part of my role as attorney general. We enforce antitrust."

The *Gazette-Mail*, one of West Virginia's most important institutions, was under siege. The newspaper wasn't just a chronicler of the state's daily events, it was a relentless watchdog. It didn't just cover stories, it uncovered them. The scrappy *Gazette* had exposed corruption and greed and incompetence. No newspaper reported on the coal industry more rigorously. The *Gazette*'s Ken Ward Jr. was widely considered the nation's best coal reporter. His doggedness led to mine safety reforms and put corrupt coal barons in prison. The paper's tenacious reporting extended to state government, schools, jails, local businesses, city hall, cops, and courts. It was an undeniable check on power.

Under local ownership, the *Gazette* and *Daily Mail* were struggling to weather the storm of downsizing that afflicted the newspaper industry nationwide. But West Virginians still had an appetite for local news, and the *Gazette* and *Daily Mail* gave it to them in spades. The combined papers would be stronger as one. And that gave Morrisey fits. The *Gazette-Mail* was the only newspaper in the state reporting on his financial connections to the pharmaceutical industry.

Morrisey's subpoena directed us to hand over balance sheets, income statements, cash flow statements, audits, a list of company officers and employees, and documents that would show the newspaper's advertisers and subscribers. It also demanded a list of operating expenses, marketing plans, circulation figures, an organizational chart, and all emails and texts about the decision to combine the papers. What's more, Morrisey ordered *Gazette-Mail* management to answer a dozen questions. "Describe all monetary and non-monetary value paid by the *Gazette* for the *Daily Mail*" was an example. Morrisey gave the *Gazette-Mail* ten days to answer his questions and fork over the records.

The newspaper's lawyers objected and sent a letter to Morrisey's office. Morrisey had, it was noted, received no complaints about the merger. And he hadn't conducted an investigation before issuing the subpoena, or if he had, the inquiry could have lasted only two days. State law required a finding of "probable cause" that an offense was being committed before Morrisey could trot out an investigative subpoena. Mere suspicion wasn't enough. There had to be more. The *Gazette-Mail*'s lawyers contended nothing was there. The subpoena was "invalid and unenforceable," the lawyers wrote. The *Gazette-Mail* refused to answer Morrisey's questions, and he wasn't getting his hands on the newspaper's financial statements. Not a chance.

Two weeks later, Morrisey filed a petition in court to enforce the subpoena and block the merger. The attorney general alleged a conspiracy to put the *Daily Mail* out of business and announced he had found probable cause that combining the newspapers "may violate the West Virginia Antitrust Act." He didn't file in Kanawha County Circuit Court in Charleston, where his office is located. Instead, he took the dispute just west across the county line to Putnam County, where both newspapers circulated and which was dominated by Republican politicians. Morrisey figured he'd be on friendlier turf there.

At a court hearing, Putnam circuit judge Phillip Stowers had questions about the large number of items in Morrisey's subpoena. The judge instructed both sides to meet. Would the attorney general whittle down the list? Would the newspaper hand over some documents willingly? Stowers encouraged compromise.

Richard Neely, a seventy-four-year-old former state Supreme Court justice who was representing the newspaper, told the judge a newspaper executive—Betty Chilton's son-in-law, Norman "Trip" Shumate—would answer Morrisey's questions. But Neely, who also was Chilton's personal lawyer, griped that Morrisey was demanding too many records: "They're asking us to produce huge amounts of documents at enormous expense that don't tell anybody anything."

Morrisey had accomplished his goal: He was back on the offensive.

The newsroom was reeling at the time, and Susan Chilton, who had no newsroom-management or journalism experience, had been anointed captain of the sinking ship. *Gazette* and *Daily Mail* reporters and editors had to reapply for their jobs. We had to submit résumés, dress up for formal interviews. Only so many slots were to be filled at the combined paper. There would be cuts. Reassignments. My job at the capitol wasn't safe. Five reporters were competing for two spots at the statehouse bureau. It reminded us of *The Hunger Games*, pitting reporter against reporter. The odds were in no one's favor.

• • •

While waiting for my turn to explain why I should stay employed at the *Gazette-Mail*, I paced in the corridor outside the newspaper's conference room, tugging the sleeves of my navy sport coat. I'd have twenty minutes to make my case. Reporters adopted different strategies to make them stand out from the pack. Some hustled to churn out more stories, to raise their byline count, to become indispensable. Others played it cool, opting to show confidence, not desperation. They kept their heads down. I tried to incorporate both strategies because I wasn't sure how to play it. The news industry was shrinking. Reporting jobs were getting harder to find by the day. I didn't want to leave West Virginia. Neither did my wife, Lori. She had founded a successful nonprofit reading center for dyslexic children—the only one of its kind in the state. Students whom the schools had given up on found a place where they shone as never before. The program changed lives.

I was told not to worry. I had reported for the *Gazette* for seventeen years, paid my dues. The *Daily Mail*'s statehouse reporters were outstanding—young, eager. They hustled. And the *Gazette*'s de facto bureau chief, Phil Kabler, was a prolific writer who knew more about the state budget and legislative procedures than anyone else in the building. Phil also wrote a popular twice-a-week column about politics. He had been a fixture at the capitol for two decades. But experience didn't guarantee you a job. The newspaper offered severance packages—one week of pay for each year of service—and a few newsroom employees took them, but not enough to stave off cuts.

We had to submit résumés and cover letters to a three-member hiring committee—Susan Chilton, *Gazette* editor Rob Byers, and *Daily Mail* editor Brad McElhinny. I hadn't updated my résumé in more than a decade. I had to call a former *Gazette* colleague, Amy Julia Harris, who was working in California, to find out how to format one. I used the cover letter to talk about the importance of investigative reporting. I cited a series I cowrote on long bus rides for children that led to school transportation reforms and ultimately reduced the amount of time students spent in transit. I spotlighted articles about the state's methamphetamine-lab problem that prompted retailers to stop selling a cold medication used to make the illegal drug. I emphasized my versatility, my ability to produce nuanced, breaking news stories and in-depth pieces. I tossed in a reference to sustained outrage.

Now, I was next up to interview. The door swung open. I took a deep breath. There was no time for coffee and doughnuts and chitchat. Susan and Rob commented on the length of my résumé. Four pages, in hindsight, was overkill. Brad asked about stories I was working on, and before I knew it, the interview was over. Two weeks later, an email popped up from Human Resources: "I am pleased to offer you the position of 'Reporter Statehouse' in the *Gazette-Mail*." We had until midnight the next day to accept or decline. I decided to stay put and stick it out.

Other newsroom employees weren't as fortunate. A longtime *Daily Mail* editor and photographer were let go. The *Daily Mail* lost its police reporter. And in a terrible blow to newsroom morale, the *Gazette*'s veteran investigative reporter Paul Nyden, who had worked forty years at the newspaper, wasn't hired back. Nyden had taken young, enterprising reporters under his wing, and he and his wife cooked supper for them at their house on Sundays. Many of the *Daily Mail*'s reporters were reassigned to beats they didn't want. They started searching for new jobs. I wondered if any would go to work for the attorney general's office. Morrisey, who was still trying to scuttle the merger, had hired away *Daily Mail* reporters and editors in the past by offering higher salaries. Now, they might have a whole new reason to jump ship: an animus shared with Morrisey toward the *Gazette*.

Momentum was shifting in Morrisey's favor. With his onerous subpoena petition still pending in one county court, he beat us in another. Kanawha County judge Charles King had issued a ruling on our public-records lawsuit seeking the email message and other correspondence that Morrisey's office refused to release.

I was finishing up lunch at Burger King—BK Veggie (no ketchup, no mayo, add mustard), fries, large iced tea—when I spotted a text on my flip phone from the *Gazette-Mail*'s city editor, Greg Moore. He asked whether I could write a story about King's decision for the next day's paper.

I texted back, "How did he rule?"

"Against us," Greg responded.

Four months had passed since the hearing. It was the middle of September 2015. We had no heads-up that King would issue a decision that day. I drove back to the capitol pressroom to write the article that no reporter wants to write. We had challenged a public official's decision in court, and we had lost. Morrisey would not have to release the email and other documents that we thought could shed light on his role in the lawsuit against Cardinal Health. What's worse, King's ruling was exactly the same as the proposed order that Tseytlin had submitted to the court months earlier. The judge made no changes. The order claimed that the email contained only a single reference to the Cardinal lawsuit, but did not show whether Morrisey took part in the case. It also contradicted statements in Tseytlin's previous filings. Morrisey was mentioned just once in the email, according to King, though Tseytlin had acknowledged two mentions of his boss. King wrote that Morrisey could keep the email secret because it was "clearly protected by attorney-client privilege." Morrisey crowed that he had defeated the *Gazette-Mail.* He said, "From the beginning, this case has been politically motivated and an unnecessary burden on this office and a drain of taxpayers' resources."

The loss stung the newsroom. We figured this would only embolden Morrisey's push to seize our private financial records and block the combination of the two papers. Everybody braced for the worst—except Pat McGinley. He was already talking about filing an appeal to the West Virginia Supreme Court.

Two weeks later, Susan Chilton, Byers, McGinley, and I met in a vacant newsroom office. A large portrait of Ned Chilton hung on the wall. McGinley explained that King's order contained factual errors. Those alone would be grounds for an appeal. We had thirty days to decide. McGinley would just need the go-ahead. He would continue to work for free. Chilton had reservations. Did we want to anger Morrisey while he was hell-bent on stopping the merger? Then Susan went further: "What if we stop covering him?"

I wasn't sure what she meant by that. The room was silent.

Susan explained that Morrisey thrived on attention. "Wouldn't it drive him nuts if we just ignored him? He'd go crazy."

I didn't know what to say. Nor did McGinley or Byers. I had spent the past two years writing dozens of stories about Morrisey and Cardinal, about his gutting of the consumer protection division, about his disregard for his own office policies, about his reluctance to sue corporations, and now we were going to stop covering his actions altogether? There would be no appeal to the Supreme Court, the Chiltons later decided.

After returning to the capitol, I called the county commissioner who helped identify Morrisey's connections to Cardinal back in 2013. I wanted to vent. I wanted advice on what to do next. I was frustrated, upset, deflated. He asked me if I needed help finding a new job.

13

The Drop

My nineteen-year-old son, Toby, never got a good look at the stranger who sneaked up to our porch and dropped something into the mailbox beside the front door. It was a weekday morning in the middle of October 2015, three weeks after the judge ruled that Morrisey could keep the email suppressed. Toby texted me at the capitol that someone had parked on our street across from the house, left the engine running, and dashed up the sidewalk to deliver a sealed nine-by-twelve manila envelope with my name on it. There was no postage, no return address, no sender's name.

During my seventeen years as a reporter at the *Gazette*, I had received dozens of anonymous letters and packages. Some led to front-page stories. Others I tossed in the recycling bin—usually letters that started with a series of questions that the tipster most likely already knew the answers to, so why not just say it? This was different. It wasn't dropped off at the newsroom. It hadn't been slipped under the pressroom door at the capitol. It was hand-delivered to my house in Charleston. Someone would have had to search public records or a phone book for my home address. My son heard footsteps and the clank of the metal mailbox lid, but by the time he parted the blinds of his upstairs bedroom window, the courier was already driving away.

"Dad, somebody dropped something off at the house for you," the text said.

"Just leave it there," I responded.

When I arrived home, the envelope was still protruding from the mailbox attached to the front of our redbrick house. On the outside in black ink ERIC EYRE was written, along with HAND DELIVERY and PERSONAL in the bottom left corner. I tore it open and flipped aside the two-prong fastener. Inside was a single sheet of paper—the email Morrisey had fought to keep secret for two years. The header indicated it had been sent at 4:45 p.m. on Monday, May 6, 2013, under the subject line "RE: State Pill Mills Cases." The sender was Dan Greear. The recipient was James M. Cagle. Greear had sent an update to Cagle for his work on the lawsuits against Cardinal Health, AmerisourceBergen, and other prescription drug distributors.

For two years, Morrisey stuck to the same story: He had stepped aside from the Cardinal Health lawsuit the first day he took office. He repeated that assertion at three public events. But here was Greear's email, unveiled, for all to see: Morrisey had given "specific instructions" for the lawsuits against Cardinal and the other drug distributors. The message referenced Cardinal by name, but no other drug company. Greear told Cagle that he had spoken to Charleston lawyer Mark Carter, who represented Cardinal and had also headed Morrisey's campaign transition team. Morrisey had directed Greear to "continue talks with defendants" and to identify any drug companies that should be added or "cut loose" from the lawsuits. Cagle and Jackson were still gathering evidence that showed which distributors were shipping an excessive number of painkillers to the state, and to which pharmacies. It didn't make sense to sue companies that had few deliveries. Cagle and Jackson wanted to focus on the worst actors.

In the email, Greear also notified Cagle that Morrisey wanted to attend a court hearing on the Cardinal case in Boone County that week, but it got postponed because Morrisey couldn't be there. Greear warned that the drug companies might try to disqualify Cagle as a special assistant attorney general if Morrisey wasn't at the Boone County hearing. The drug distributors' lawyers had been arguing that the attorney general didn't have the right to hire outside lawyers to help sue them. Morrisey wanted to make clear he was calling the shots. He would determine the fate of the lawsuits against the distributors.

On May 8, which was two days later, Greear and Cagle had filed a motion in Boone County Circuit Court requesting that a hearing be rescheduled. The filing, titled "State of West Virginia v. Cardinal Health," which mentioned Morrisey by name, showed that Greear and Cagle had followed Morrisey's instructions and declared that the hearing couldn't take place without Morrisey being in the courtroom—all at a time when Morrisey would later say he had no role in the case. In reality, Morrisey wanted to run the show, to send a message that he was in charge of the lawsuits against the distributors. Now, in the fall of 2015, after fifteen months of Morrisey's denials, the email was the evidence I needed to show that Morrisey had been involved in the lawsuit against Cardinal at the same time his wife was lobbying for the company. I started to plan how to use the leaked email.

In the copy I had, I noticed the message's body and header incorporated two distinct typefaces. It appeared someone had cut and pasted the two sections together. I wondered whether the email had been doctored, or perhaps it was a fake. But it wasn't. I read the email over the phone to a lawyer who was in a position to confirm its authenticity. Now, all I had to do was figure out what it all meant, but I had no assurance the newspaper's management would even give me the green light to publish a story in the face of threats from the attorney general. That decision would have to come from the top.

Morrisey didn't know I had Greear's email, but he quickly found out. I sent a list of questions to Morrisey, asking him about the email and the court filing. Morrisey denied that he had given

Greear any instructions about the Cardinal lawsuit. He insisted that he issued instructions only for the second lawsuit against drug distributors. He passed blame to Greear. "Any statement by anyone suggesting otherwise would be contrary to my explicit instructions," Morrisey said.

The attorney general wouldn't comment on the rescheduling of the Cardinal court hearing or why he had to be there. I had hunted down that filing after spotting a reference to it in Greear's email. I inquired why Morrisey's office didn't release it in response to my public records requests—or even acknowledge that it existed during the court proceedings before Judge King in Charleston. One of Morrisey's top aides, Anthony Martin, responded that I had requested *emails* about Morrisey's involvement in the Cardinal lawsuit; the court filing that announced Morrisey's unavailability for the hearing existed in the attorney general's records as an attachment to a May 8, 2013, email, and I had not asked for attachments. (In all my years as a reporter in West Virginia, this was the first time a state agency had made that distinction. Other agencies considered an attachment to be part of an email.)

Martin wrote to me, "The filing was not subject to your prior Freedom of Information Act request because that FOIA was limited only to 'emails.'" That afternoon, Martin marched down from his office at the state capitol and knocked on the pressroom door. He wanted to talk. I was on the phone. By the time I stood up and opened the door, he was gone. Minutes later he emailed me to say he was consulting with Misha Tseytlin on the "legal implications" of our plans to publish a story about Greear's email to Cagle, and how it contradicted Morrisey's insistence that he had stayed out of the state's lawsuit against Cardinal since the day he took office.

The following day, Misha Tseytlin was on the phone with Pat McGinley. Tseytlin was threatening "legal action" and "court-imposed sanctions" against the newspaper if we published a story that quoted the leaked email that had found its way to my front porch. Judge King had placed Greear's email under seal, Tseytlin reminded McGinley. And now the *Gazette-Mail* was going to violate a court order. Didn't McGinley understand there would be serious consequences?

"Pentagon Papers," McGinley said. "I assume you're familiar with that case?"

McGinley was citing the landmark Supreme Court decision on the First Amendment that made it possible for the *New York Times* and the *Washington Post* to publish classified documents without risk of government censorship or sanctions. President Richard Nixon had ordered the *Times* to suspend publication of the papers, which detailed top-secret US actions in the Vietnam War.

Of course, Tseytlin was familiar with the case. He'd graduated from Georgetown University, had clerked for Justice Kennedy, and was a frequent speaker for the Federalist Society. He knew about the Pentagon Papers.

McGinley, who was attending a conference in Tennessee that day, suggested that Morrisey might want to rethink his plans to block publication of the story. McGinley asked Tseytlin to put his objections in writing. "You're surely aware that would be an unconstitutional prior restraint that would violate the First Amendment guarantee of freedom of the press," McGinley said.

As I was finishing up the story late Saturday afternoon, Tseytlin's written warning popped up in my in-box, forwarded to me by McGinley: "It is deeply disturbing that your client appears willing to publish a document that is protected by the attorney-client privilege and sealed by court order. I urge your client not to violate the court order, risk sanctions, and, in the words of Judge King, compromise the state's vital interests."

McGinley figured Morrisey was bluffing, but the warning was stirring unease in the newsroom. Susan Chilton wanted to be included in the discussions about the story and any last-minute fireworks. The editors instructed me to send her a copy of the article before it went to press. This unusual request gave me pause. Morrisey was investigating the *Gazette-Mail* merger. He wasn't interested in compromise, as the judge had directed. He had the newspaper on the ropes. I worried the story might get spiked to appease him.

At 5:30 p.m., McGinley reviewed my story, spoke to Chilton and executive editor Rob Byers on the phone, then followed with a memo. McGinley concluded that Morrisey had no legitimate grounds for sanctioning the newspaper for writing about Greear's email. Chilton gave the go-ahead to publish. Then she went a step further. I was at my desk at the capitol the next morning when McGinley called to tell me about her follow-up email to him—a thank-you note. I knew immediately this was more than a gesture; it was a turning point, a statement made as powerful forces were closing in to smother her family's livelihood and legacy. She had moved her chips to the middle of the table. She was betting on sustained outrage.

"Your guidance certainly helped me through my first difficult decision of whether or not to run a story," Chilton wrote. "I felt much more confident after having spoken with you and reading your communication concerning our story. You reminded me of how much a community depends on its newspaper to tell the truth and follow through finding the truth even if it's a little scary."

For his part, Morrisey was still insisting the email didn't say what it did and that Greear was mistaken for thinking he had to reschedule a Cardinal court hearing because the attorney general couldn't make it. Morrisey denied that he wanted to control the case. In a couple of weeks, his lawyers would be back in court in Putnam County, pressing the judge to force the newspaper to turn over its books and stop the merger.

On the first Thursday of November 2015, Judge Phillip Stowers announced he was going to rule from the bench on Morrisey's petition to subpoena records that would detail the newspaper's financial troubles and give him ammunition to block the merger. Morrisey was hornet mad about a story I'd written two weeks earlier—and it had nothing to do with Cardinal Health. He had hired a veteran political strategist and put him on the state payroll as deputy chief of staff. Though the new hire had never practiced law, he was going to tell attorneys in Morrisey's office what to do, and his $95,000 salary was being paid out of Morrisey's consumer protection fund. Newspapers and television and radio stations across the state picked up the article. Morrisey responded that my report was riddled with errors. Two days later, the political operative resigned.

In Stowers's courtroom that day, the Chilton family lawyers argued that Morrisey was simply out to punish the newspaper because it exposed his "nefarious dealings." It would cost the *Gazette-Mail* thousands of dollars to compile the documents that Morrisey subpoenaed. He alleged there was a conspiracy to shutter the *Daily Mail*, but the Chilton family owned both newspapers, and you can't conspire with yourself.

"Just because someone was in prison once doesn't mean they will automatically break the law again," Stowers interjected.

The judge had more points to make. Morrisey's office hadn't fielded any complaints about the merger. Subscribers weren't griping. Competing media outlets weren't upset. The Feds weren't challenging the decision. So why was Morrisey pushing so hard for a subpoena? Why was he insisting the newspaper was violating state antitrust laws?

The judge told the lawyers he was dismissing Morrisey's petition. Stowers said, "The chilling effect of allowing a state actor, whether it's the attorney general or another state entity, to be able to, on mere supposition, require a person involved in First Amendment activity to have to spend a bunch of money just because someone thought it was a good idea to look into it, that's not what this statute envisioned."

Morrisey's assistant attorney general refused to comment after the proceedings. A spokesman said Morrisey was evaluating next steps. But Morrisey did not appeal Stowers's decision. He had other things on his mind. He had a reelection campaign to organize—and the lawsuits against the drug distributors to settle.

14

Dragging and Lagging

In mid-December 2015, West Virginia's lawyer disciplinary board quietly concluded an investigation of Morrisey and dismissed an ethics complaint against him. The investigative panel forwarded a copy of the decision to the attorney general's office, and there it sat. Though the panel, as a matter of policy, did not publicly announce its decisions, Morrisey could have put out a release, held a press conference, declared victory. But he didn't say a word. The seventeen-page report disclosed his sworn testimony, and that would require a lot of explaining.

The investigation had been kept confidential. I obtained a copy of the investigative panel's dismissal report three weeks after it was issued. It let Morrisey off the hook, but not before calling out his conduct. In interviews with the panel, Morrisey acknowledged he didn't "permanently screen" himself from the Cardinal lawsuit until July 2013—after I reported that the company had helped pay for his inaugural party and employed his wife as a lobbyist. Morrisey also revealed to the panel—and this was news to me—that he represented Cardinal on legal matters before being elected attorney general. He gave no specifics. The report concluded that Morrisey provided "directions" to Greear about how to supervise Cagle, attended a closed-door meeting with Cardinal's lawyers, and asked that the Boone County court hearing about the case be rescheduled because he couldn't attend, as the Greear email anonymously delivered to me had already shown. But none of those actions violated the lawyers' rules of professional conduct. And Greear testified that Morrisey had never pressured him to drop the lawsuit or settle for a low price. The suits against Cardinal and AmerisourceBergen and other wholesale drug distributors were proceeding, albeit slowly.

Morrisey told the panel that his wife's lobbying for Cardinal didn't pose a conflict of interest—and that he could have continued to take part in the lawsuit. In a sworn affidavit provided to the disciplinary board, he asserted that he had "no material interest" in any compensation his wife received from Cardinal. Her income had "no impact" on his judgment. The panel, however, found the circumstances created an "appearance of impropriety," a bad look considering Morrisey was the state's chief legal officer. Shouldn't lawyers who hold public office be held to a higher standard? The panel sure thought so. While it didn't charge Morrisey with any ethical violations, it put him on notice: "The attorney general is strongly warned regarding his duties pursuant to the conflict of interest provisions of the rules of professional conduct, and warned that any future violations of the rules may result in sanctions."

Morrisey was on a short leash. He needed a distraction. He had to show he wasn't complicit with the distributors. Just hours after I contacted his office for comment on the investigative panel's report, Morrisey announced he was suing McKesson Corporation, the nation's largest prescription drug distributor, for allegedly fueling West Virginia's opioid crisis.

Starting in August 2014, Governor Earl Ray Tomblin's cabinet secretaries had been pressing Morrisey to sue McKesson. Karen Bowling, who headed the Department of Health and Human Resources, and Joe Thornton, chief of the Department of Military Affairs and Public Safety, wrote letters to the attorney general.

Even earlier, by the first week of January that year, Cagle and

Jackson had the evidence they needed to add McKesson as a defendant in West Virginia's lawsuit against the distributors, but Morrisey's assistant attorney general Maryclaire Akers told them he had other plans for McKesson, and under no circumstances were they to sue the company.

Lawyer ethics rules had stopped McGraw from suing McKesson when they filed the original lawsuits eighteen months earlier. The American Bar Association's Rules of Professional Conduct, No. 11, requires lawyers to have a factual and good-faith legal basis for filing a claim, and Cagle and Jackson had no records that showed McKesson shipped prescription drugs to pharmacies in West Virginia. That changed the following year, however. Sav-Rite Pharmacy's lawyer found two boxes of invoices. He was duty bound to turn them over to Cagle as part of the ongoing lawsuit filed by recovering addicts.

Cagle's paralegal drove to a Charleston law firm to fetch the documents. "They're McKesson," she said, plopping the boxes down on Cagle's cluttered conference table. Of the hundreds of McKesson invoices for drugs sold to Sav-Rite, most were pain pill orders.

"Bingo," Cagle said.

Around the same time, Cagle was drafting updated lawsuit complaints against the drug distributors. For the state, he added McKesson's name as a defendant in the lawsuit against AmerisourceBergen and the eleven smaller companies. He cited McKesson's shipments to Kermit in the amended complaint.

But those additions never made it into the final draft. Akers, on her boss's orders, directed Cagle to remove McKesson, and if Cagle and Jackson wanted to stay on the Cardinal and Amerisource Bergen case, well then, they had no other option but to do as told. She would explain things later, she said. So Cagle filed the updated suit against AmerisourceBergen in Boone County Circuit Court without adding McKesson as a defendant.

In the months that followed, Morrisey's office wouldn't

respond to the governor's cabinet secretaries' pleas to sue McKesson. Bowling and Thornton had already signed their agencies onto the Cardinal and AmerisourceBergen lawsuits. They wanted to recoup some of the hundreds of millions of dollars their agencies were spending on problems caused by prescription drug abuse—addiction treatment, health care, foster care, and prison administration. Criminal investigations and prosecutions. Body transports following overdoses. Autopsies. Bowling and Thornton were getting frustrated by Morrisey's dodges and delays. How difficult was it to append just one company to the list of distributors being sued? Why was it taking him so long?

A month later, I asked Morrisey's office those questions, and once I did, Morrisey moved swiftly to dispel any notion he was dragging his feet. I had reported that, before taking office in 2013, Morrisey had lobbied for the Healthcare Distribution Management Association, or HDMA, the trade group that represented McKesson, Cardinal, AmerisourceBergen, and other distributors. Doing so generated $250,000 for his DC law firm. Morrisey had also attended the private meeting with McKesson executives back in May 2013—the one he wrote about in his commentary published in the *Daily Mail*.

Three days after I made calls to the attorney general's office about Morrisey's reluctance to go after McKesson, Akers filed a petition in Kanawha County Circuit Court asking a judge to force McKesson to comply with a subpoena. The filing disclosed that Akers had mailed the subpoena to the company in March 2014—after she directed Cagle to scrub McKesson from the updated suit against AmerisourceBergen. The subpoena asked McKesson to turn over records of all narcotic drug shipments to West Virginia for the previous four years. Morrisey gave McKesson thirty days. The company never responded.

Less than an hour after going to court to enforce the subpoena, Morrisey posted notice on his website that his office was soliciting bids to hire an outside law firm to investigate McKesson. The move took Bowling and Thornton by surprise. They were satisfied with Cagle's work. He and Jackson had spent two years on the lawsuits against drug distributors. Why not go with the guys who had a running start?

I contacted McKesson's spokeswoman. The company knew nothing about the subpoena or investigation, she said. I sent her postal records—Akers attached them to the petition—that showed the subpoena was delivered, and I forwarded the online posting that sought a law firm to investigate. The spokeswoman then acknowledged the subpoena and investigation. McKesson, she said, was now working with Morrisey's staff to figure out what documents the company would provide.

The investigation, however, dragged on for more than a year. Democratic state lawmakers joined the call to tag McKesson to the state's lawsuit. Morrisey responded that his now-eighteen-month probe needed more time. He had yet to determine whether McKesson broke any state laws. He had also yet to hire a law firm to investigate. "To date, the investigation has been totally in-house," a Morrisey aide told me for a story I was writing about the attorney general's delays, "so any media coverage or other information into the status of the McKesson investigation or the reasons for any actions taken by this office is nothing more than conjecture and speculation."

What wasn't up for speculation was that Jim Cagle and Rod Jackson wouldn't be representing the state against McKesson. Morrisey hired another law firm, one that had no experience suing pharmaceutical companies, to prosecute that portion of the case. Morrisey didn't like Cagle and Jackson. He didn't trust them. They weren't following his script. They were making lawyers for the drug distributors angrier by the day, threatening to reveal more secrets about their painkiller shipments.

15

A Door Cracked Open

Boone County circuit judge William "Will" Stewart Thompson sat in his chambers, thinking about Jim Cagle's request. On behalf of the State of West Virginia, Cagle had asked Thompson to toss out a court-ordered agreement that kept the distributors' pill shipments to all pharmacies in the state confidential. It was the last Monday in February 2016. Thompson's oak gavel rested atop his tidy desk. On the yellow wall behind him, above dark-stained wainscoting, was a picture of his great-great-grandfather, the county's former high sheriff. Thompson's ancestors had been living in the area since before the Civil War, before West Virginia achieved statehood in 1863.

In late November of 2013, Judge Thompson had ratified a confidentiality agreement between Morrisey's office and Ameri-sourceBergen, along with eleven smaller distributors such as Miami-Luken and H. D. Smith. The decree allowed the companies to conceal records and testimony—and perhaps entire pleadings—as the state's lawsuit against them proceeded. Cardinal Health and Morrisey's office filed an identical agreement that December.

The secrecy deals permitted the companies to label documents confidential or highly confidential. The drug distributors weren't going to take any chances. They wanted to keep certain things out of the court file, information that the citizens of West

Virginia didn't need to know. The agreements, called protective orders, were a way to keep information under wraps, away from reporters that might come snooping. The agreements also mandated that the state return the confidential records after the cases closed. That ensured the information would stay secret forever.

By early 2016, the drug distribution companies had turned over their shipping records, tens of thousands of pages, to Morrisey's office, and by extension, to Cagle. The records included drug prices and delivery dates. The companies had marked the information highly confidential. The last thing they wanted was for the whole world to see it.

But now Cagle was asking Thompson to blow up the secrecy deal. Cagle wanted details of the distributors' pill shipments to get out. Morrisey could do little to stop him—a month earlier, the attorney general announced he was stepping aside from the state's lawsuit against AmerisourceBergen and the regional distributors (along with his now-official recusal from the Cardinal case). Cagle was fed up with the companies' delay tactics. The lawsuits had puttered along for more than three and a half years. He wanted to get them moving again. This would get the distributors' attention. Only Thompson stood in the way. The companies had urged the judge to keep the 2013 confidentiality agreements intact.

In a few hours, Thompson would deliver his decision to his law clerk, who would type it and see that it got filed before the end of the day. The moment weighed heavily on the judge. Private businesses had the right to keep proprietary information private. But Boone County residents needed to understand the affliction that gripped their community, and Boone County was home.

Thompson had returned to Madison, the county seat, after graduating from West Virginia University law school in 1995. As a young lawyer, he mostly handled child abuse and neglect cases—about forty a year. He'd advocated for children yanked from their homes because they had chronic head lice or lacked clean drinking water. In juvenile court, he had represented boys arrested for throwing rocks through windows. He had a case where a child stole a teddy bear.

In 2007, Governor Joe Manchin appointed Thompson circuit judge in Boone and Lincoln counties. At thirty-seven, he was the youngest circuit judge in West Virginia. He filled the post as the opioid problem worsened. Child abuse cases increased fourfold. Nearly all were drug related. Mothers and fathers chose pills over parenting. They didn't care about getting a job or making sure the kids got fed and went to school.

Thompson remembered an eighth-grade girl who landed in juvenile court. She was skipping school. The county had a take-home backpack meal program. But since she was missing school, she wasn't getting the meals. She lost ten pounds in a week. Thompson ordered her to be taken away from her mother. The girl protested.

"How could I send you back if your mama doesn't feed you?" Thompson asked.

"I don't eat very much."

Most parents didn't bother showing up in Thompson's courtroom to fight for their children. Let the state have them. Just get me my next fix. Some eighty hearing notices would go out, only a dozen parents would make their way to court, and it would be a good day if one was doing things right.

Parents wouldn't even take their children to the dentist to soothe a toothache. Thompson had to order them to do so. "If you don't do it, I'm going to hold you in contempt of court, and I'm going to put you in jail," the judge told one mother. And he did exactly that. Her addiction smothered everything else, extinguished her motherly instinct to help her child in pain.

"My teeth don't hurt. My teeth don't hurt," the boy screamed as the deputy led his mother out of the courtroom.

"Do your teeth hurt?" Thompson asked the boy after she left.

"Yeah, they really hurt."

The judge had his own children—three daughters and a son—

to worry about, too. He was a father and a husband first. And damn if he was going to let pills rot his community. Folks needed to know what they were dealing with, what was driving the decay they saw firsthand every day. The levy had burst and poison pooled in the streets. All of a sudden homeless people were stumbling around Madison like zombies. In the close-knit town of twenty-eight hundred people, neighbors were having to police the parents who showed up high at softball games and to pick up needles in the outfield. How do you explain that to your daughters? Thompson wondered. What do you tell your son when he asks about the homeless encampment under the bridge after he paddles his kayak down the Little Coal River? The funerals mounted. Courthouse employees were laying to rest their loved ones. Thompson's own neighbors were burying their children who overdosed. When he spoke at community forums, he opened with a dire warning: "If you can't say your family's been affected by drug abuse, substance abuse, you just don't know your family very well."

At first, Thompson considered drug addiction a choice. People chose to take pills. People chose to get high. People chose to become addicts. But over time, after overseeing thousands of drug cases, he came to see it as a disease. His courtroom would fill up with drug abusers who believed they could stop when they wanted to. But they didn't. They needed intervention, drug courts, addiction treatment. He gave them second and third chances to turn their lives around. For those who did, they came to view Thompson's black robe more as a crusader's cape—he fought for them, believed in them, hoped for them.

In 2012, McGraw, the former attorney general, could have filed lawsuits against the drug distributors anywhere in the state. But he picked Boone County, a place hit hard by the addiction crisis, a place with a no-nonsense judge who lived the crisis day after day in his courtroom. Thompson had a spine. He wouldn't cower to corporations and their well-compensated attorneys. But he wouldn't roll over for the trial lawyers either.

Back in his office, Thompson wasn't going to make his law clerk wait any longer. He had made up his mind about Cagle's request to void the confidentiality deals, which would in turn make public the revised lawsuits Cagle had recently filed against the distributors. Those revisions added information Cagle had culled from their shipping records.

After making final edits to his ruling, Thompson stood up from his desk, slipped on his black robe, grabbed his gavel, and spoke to his clerk on their way to the courtroom for an unrelated hearing. He had grappled with his decision, weighed down by his experiences, but he hadn't heard enough to void the court order that protected the confidential records. They would stay under lock and key—for now. He instructed his clerk to file a court order denying Cagle's request. But the judge left the door open, as best he could, to reconsider his decision. He asked both sides to submit proposals for new ways of handling confidential documents in the future. Maybe portions of filings could be unsealed instead of its being all or nothing. Thompson seemed to signal that he might rule differently if the request to open the court records came from someone not subject to the confidentiality agreement, someone who could argue that the public had the right to know the details of a lawsuit filed by an elected official—the attorney general—on behalf of the state's citizens.

Later that week, I called the courthouse records room. A clerk answered the phone. I asked for a copy of the lawsuit's docket, which listed all filings in the AmerisourceBergen case since day one. I wanted to see the entry for Cagle's latest filing of the revised lawsuit against the distributor and confirm the date. A dusty fax machine at the capitol pressroom spat out the docket sheets one at a time. I traced my index finger down the pages, stopping on line 172, confirming that the court records remained "FILED UNDER SEAL."

Just then, my phone rang. Rod Jackson wanted to give me a

heads-up about Thompson's latest ruling. I had been knee-deep in covering the closing weeks of the legislative session. Jackson suggested the newspaper might have a shot of swaying the judge to unseal the court files. The press could argue that the public's right to know outweighed a corporation's interest in keeping information secret. I was listening. If we needed a local attorney to assist Pat McGinley, Jackson and Cagle had someone in mind—a lawyer who knew Cagle from their days together as legal aid attorneys representing the poor in southern West Virginia, a lawyer who wouldn't be hard to find. Tim Conaway's office was directly across the street from the courthouse in Boone County.

PART III

16

Eighteen Words

Pat McGinley and I didn't realize the magnitude of what we were up against until we pushed through the wooden double doors of the courtroom and glanced at the brigade of lawyers crowding the first two rows of benches. The drug distribution companies were paying them $500 and up per hour to stop us from getting our hands on court records that detailed pill shipments to West Virginia. They were used to winning.

I was there, in the middle of April 2016, to report on my employer's decision to go to court to unseal the state's lawsuit against the drug distributors and persuade the judge to reverse the previous ruling keeping their records secret. After the *Gazette-Mail* filed a motion to intervene in the state case, we were at a tipping point, ever so close to unveiling evidence about the origins of the addiction crisis, a man-made disaster fueled by corporate greed and corruption.

The Boone County Courthouse dated back to 1921, and its lone courtroom had changed little over the years. A central skylight glowed with sunshine. Brass coat hooks studded the walls. The floorboards creaked. In the spectator gallery, a balcony loomed over rows of dark-stained benches. Directly over the judge's bench was a second balcony, the high ground above decades-old battles, the perch where deputies stood, armed with rifles, watching for trouble back when trials pitted union mine workers against coal companies.

On this day, the dozen attorneys representing the drug distributors—ten men, two women—had traveled from Pittsburgh, Philadelphia, Charleston, and Columbus, Ohio. Cardinal Health's lawyer would introduce himself later as Jim Wooley, no relation to the pharmacist from Kermit, but a strange coincidence. Jim Cagle and Rod Jackson, who remained on the case as outside lawyers representing the state, took seats at a table up front beside the jury box. Two lawyers from the attorney general's office kept them company. McGinley, who had the day off from teaching at the law school in Morgantown, chatted with our cocounsel, Tim Conaway, a coal-country lawyer who, with his son, shared a law office, Conaway & Conaway. Tim understood what prescription drugs had done to his county, which had one of the highest overdose death rates in the nation. He also knew Judge Will Thompson.

"All rise," the bailiff thundered.

The judge called on the lawyers for the drug distributors. Two jumped up. They urged Thompson to postpone the hearing. Some companies, they said, were close to being dismissed from the lawsuit. So why should anything they've done be made public?

"All right, I think I understand your position," Thompson said. "Mr. Conaway, Mr. McGinley, tell me why I should hear this today."

Conaway went first. "Well, Your Honor, we think not allowing this information out is harmful to the public. The sooner this information is out, the better."

It was extremely rare for a lawsuit complaint to be sealed, especially one filed by the attorney general on behalf of West Virginia's citizens. When McGinley got a turn, he tried to eviscerate the distributors' contention that court documents ought to remain sealed because they didn't want their competitors to get a peek at their sales figures.

"We don't know what's in those records, how many doses of addictive drugs were sold in West Virginia—how would that affect them?" he said. "There is nothing in the record to support that allegation."

The distributors had selected Alvin "Al" Emch to rebut McGinley's arguments. Emch was wearing a pinstripe suit, a pocket square flowering from his breast pocket. He was always the smartest guy in the room, never short for words. He had been fighting the state's lawsuit since almost the day it was filed in July 2012. Emch reminded the judge of the court-ordered agreement to keep confidential records private: "The one thing I will reiterate is, the court has procedures in place that the court is presently following."

In court filings the previous week, the drug distributors went after the *Gazette-Mail*. The companies claimed the information contained in the sealed documents was available elsewhere— from the DEA and West Virginia Board of Pharmacy. The DEA had the pill numbers, but it refused to release them. The pharmacy board didn't track distributors' sales. It only reported statewide pill counts each year culled from doctors' prescriptions. Emch wrote, "And if those agencies protect any of that information from the intrusive journalistic nose of the *Gazette-Mail*, then its confidential nature must be respected." *Intrusive journalistic nose* wasn't meant as a compliment, but our editors and reporters seized on the phrase in Twitter postings anyway. In our minds, it was a badge of honor. Emch called our efforts to open the sealed documents premature, unnecessary, and ill-advised. He wanted Thompson to kick the press and public out of the courtroom during any discussions about confidential records.

"The bottom line is that this is civil litigation," Emch argued. "It is not automatically an open book that the *Gazette-Mail* or any other member of the public is permitted to review."

"Does anyone else have anything to add?" Judge Thompson asked. He promised that he would rule on our request to unseal the court documents within a week. He hinted that some of the smaller drug distributors were close to reaching settlements, and he asked them to let him know soon whether they would object to the state's allegations against them being made public. The regional distributors were willing to pay small fines to keep their shipments confidential. And the settlements wouldn't require them to admit guilt.

"I want to do this rather quickly," Thompson said. "I do think this is an important issue."

The judge had one last thing on his mind. A producer from CBS News had recently called him about a conversation she had had with Morrisey's spokesman about the ongoing lawsuits. The spokesman had refused to comment, blaming Thompson for restricting any discussion of the case outside the court. The producer wanted an explanation for the gag order. Was Morrisey trying to hide something, or was it just a matter of confusion?

"I have not entered a gag order in this case," Thompson informed the lawyers.

At a previous hearing, lawyers for the distributors had griped about Cagle's giving an interview to CBS, and Thompson seemed sympathetic with the companies then, but he made clear now that he hadn't forbidden any comment at all.

"I indicated that I wanted cases tried in the courtroom, not in the media," Thompson said. "I will remind the parties that picking a jury in this, whether we do it in October or . . . next year, it's still going to be very difficult."

The judge expected it would take at least three days to empanel a jury. He planned to question potential jurors extensively. "So take that as words to the wise," he said.

When Thompson's ruling finally came down, he ordered that the state's suit be unsealed, but he gave the distributors fourteen days to appeal to the West Virginia Supreme Court before anything would be released. He concluded that the companies were trying to keep a lid on embarrassing information and protect their corporate images.

Instead of an appeal, however, the distributors suggested a curious alternative. Fine, go ahead, release the records, they said, but keep eighteen words secret. Just black them out. Eighteen words—that's all we're asking. They organized a conference call with Thompson to make their case two weeks after the ruling. I was listening in on the phone at my desk at the capitol. It wasn't uncommon for Thompson to hold short hearings via conference calls. They were vehicles for expediency.

"I have forty-seven cases on my docket today," Thompson said at the start, "so y'all are going to have a very limited time. You've got about five minutes, Mr. Emch."

Emch had filed a flurry of motions in recent days. He asked Thompson to review in private the eighteen words his client sought to withhold. Take a look, then decide. But do so behind closed doors.

"All right, I will run through it very quickly," said Emch. "This dispute has gone on for an awfully long time. I have counted approximately eighteen words out of all the pages . . . that we would ask to be redacted." The sealed filing was sixty-three pages. Eighteen words didn't sound like much. Emch again cited the three-year-old court order that classified records as confidential and highly confidential. The eighteen words fit that description. "The defendants always protect their sales information," Emch said. "They do that zealously." The DEA had the numbers but protected the same information under federal law, and sharing it would raise antitrust issues, Emch alleged. It was a matter of business principles, integrity, and maintaining healthy competition.

"Our effort is not a concern about corporate image," he continued. "This is business, confidential proprietary data that's provided to the DEA under protections against disclosure. No good reason exists at this point to disclose that information." Emch was reiterating old arguments. The ruling Thompson had put on hold for two weeks declared that the drug distributors' business interests in keeping the court documents sealed didn't outweigh the public's right to see them. Nothing was in there about prescription drug prices or profits. What was there to hide?

"Here's what I'm going to do," Thompson said. I listened on my phone at the capitol, jotting notes. "Me and my clerk are going to spend the next two hours drafting an order. I'll make my decision, and you'll have it by noon. Anything else?"

No one spoke. The hearing was over. You could hear a ding every time someone hung up. By my count, twenty-one people had been on the call.

Those eighteen words the distributors fought to keep secret weren't words, after all. They were numbers. Big numbers. Numbers of pain pills. Tim Conaway faxed me a copy of the unsealed court document after Thompson ordered its release, denying the distributors' Hail Mary. "The 18 'words' that AmerisourceBergen seeks to redact appear to be actual numbers that represent sales figures," the judge wrote.

With the numbers no longer blacked out, I could see that AmerisourceBergen alone had distributed 119 million doses of highly addictive drugs to West Virginia pharmacies between 2007 and 2012, or roughly eighty pills for every man, woman, and child in the state. About 90 million of those pills were prescription opioids such as Lortab, Vicodin, and OxyContin. The company shipped another 27.3 million tablets of alprazolam or Xanax, the antianxiety medication that addicts often took with painkillers. Even smaller wholesalers, such as H. D. Smith, had big numbers—12.4 million hydrocodone pills and 3.2 million oxycodone pills over the same years.

Hundreds of thousands of painkillers were flooding into small towns, to mom-and-pop pharmacies—some of which had filled prescriptions from doctors who later were convicted of federal crimes. H. D. Smith stood accused of selling thirty-nine thousand pain pills over two days to two pharmacies in Williamson. Another distributor, TopRx, delivered more than three hundred thousand hydrocodone pills over four years to the town of War, West Virginia, population 808. AmerisourceBergen supplied eight thousand hydrocodone pills over two days to a drive-through pharmacy in Madison. Nearby chain pharmacies weren't dispensing that many tablets in a year.

I called Don Perdue, a pharmacist and state legislator, for his reaction to the revelations. "The distribution of vast amounts of narcotics to some of the small towns and unincorporated rural areas of our state should have set off more red flags than a school of sharks at a crowded beach," he said. But it didn't. The companies and the regulatory agencies looked the other way.

By the time I received my copy, I had only an hour or two to read it over before filing a story. Entire pages were blacked out. With so much missing, it was difficult to comprehend. Thompson had redacted pill shipment information for seven smaller distributors that had agreed to settle with the state. Six of them reached deals after we filed our motion to unseal. Miami-Luken had settled the state's claims for $2.5 million three months earlier.

Lawyers and a spokesman for the drug distributors downplayed the numbers that they had fought so hard to protect for nearly four years through dodges, deflections, and delays. The totals were merely allegations, they said. It was easy to take them out of context. They made up only a small fraction of all medications shipped to pharmacies in West Virginia.

Our city editor, Greg Moore, called me to get a sense of the story's importance. I was pressed by deadline, still trying to get my head around the numbers. The conversation was short. The next day, we published the article quietly in an inside section of the newspaper. Still, it became the top trending story on our website. It stayed the top story for two weeks.

"I wish you would have told me it was a bigger story," Greg said afterward.

I shrugged. "I didn't want to oversell it."

Days later, I scoured the unsealed court records a second time.

I was working on a weekend story about the eighteen "words" being numbers. The updated lawsuit made clear that the shipment numbers to individual pharmacies were culled from tens of thousands of pages of sales information that distributors had turned over to the state. But the aggregated statewide numbers came from a different source. Buried in the complaint were references to the Drug Enforcement Administration, such as "DEA records indicate" and "the records of the DEA reflect." But how would I go about getting them?

I was aware the DEA posted some information about drug shipments on its website, but the agency listed drug quantities in grams and locations by ZIP code prefix. It was pretty much useless. In rural West Virginia, a ZIP code prefix (the first three numbers) sometimes covered five or six counties, and typically portions of different counties. Moreover, there was no breakdown by distributor or pharmacy. I considered filing a public records request with the DEA for the numbers cited in the state's unsealed lawsuit. I floated the idea to a spokeswoman at Cardinal Health. Our conversations were always cordial. She understood I was doing my job, and I understood she was doing hers. The spokeswoman reminded me that the DEA's tight-lipped reputation had earned it the nickname Don't Even Ask. But the pill numbers had to be somewhere else. I made a few phone calls. It didn't take long to find out that a second agency had possession of the same records. And I wouldn't have to go far from the capitol pressroom to get them. They were in Patrick Morrisey's office, just upstairs. But those stairs would prove a frustrating obstacle, professionally, personally, physically. I'd spend the next two months fighting for those records while my body began to fight me. Climbing those steps was a slow grind with newfound urgency. My health had put me on an accelerated deadline.

About a year earlier—or was it two years?—I noticed that my handwriting was suddenly getting smaller. My right foot started dragging. I had an unexplained cough, but only when I talked. And my right elbow—it just wouldn't straighten out. It was stiff, locked in an L. I thought it was tennis elbow. By the summer of 2016, two months after the judge unsealed the state's lawsuit against the distributors, I headed to an orthopedic surgeon—he had replaced my wife's hip a year earlier—and he ordered up an X-ray, couldn't find any tendon damage, and referred me to a neurologist. By the look on his face, I gathered something was wrong.

The neurologist hooked up a series of needlelike electrodes to my right arm. It was a Saturday morning. I was the only patient in the office. I was having an electromyography, or EMG. A machine shot bursts of electricity into my elbow. My arm twitched with each jolt. He was testing for nerve damage. I passed with flying colors. I figured I was in the clear; maybe a referral for some physical therapy for a strained elbow and I'd be on my way. But the neurologist started examining my face. Next he asked me to take a walk across the room. Only then did I realize he had used the EMG to rule out other possibilities.

"You probably have Parkinson's," he said, then handed me a prescription for a Parkinson's disease treatment drug that's also used to diagnose the disease. If it works successfully to lessen symptoms—such as stiffness and tremors—there's a good bet you have it.

I didn't want the drug to work. And it didn't. At least after I stopped taking it. It gave me a metal-like smell in my head. I couldn't sleep at night. My symptoms persisted. I started to research Parkinson's on the Internet. I read through the common symptoms: small handwriting (known as micrographia), muscle stiffness, lost sense of smell, tremor, excessive sweating, loss of balance, uneven gait. I had them all except loss of smell.

I sent an email to the founder of the West Virginia Parkinson's Support Network. George Manahan owned a marketing and public relations firm. We had found ourselves on opposite sides when my reporting prickled public officials. They would hire George's firm to do damage control. But he wasn't one to hold a grudge. George suggested that I tell my editors about my potential diagnosis right away. PD symptoms can make you look as if you were inebriated, he told me. I had some trouble walking. My wife, Lori, had noticed I slurred some words. I told the bosses. George also mentioned a new doctor at Marshall University's hospital in Huntington. He had just completed his residency. He was a movement-disorder specialist, the only one in the state. I called for an appointment. I was in his examination room the next week. Lori drove.

Dr. Vikram Shivkumar put me through a battery of tests. I closed my eyes and counted the months of the year backward. I shuffled up and down the hallway. I tapped my fingers, one at a time, to my thumbs. I squeezed his hand. He typed something into the computer, then swiveled around on a stool to deliver the news. Only he started coughing. He was having a coughing fit.

"Excuse me, I need to get some water." He left the room.

We waited.

He returned with a bottle of water, screwing the cap back on. "You definitely have Parkinson's."

Lori cried. I said nothing. It wasn't a surprise. I had done the research, done the reporting. I knew this relentlessly progressive disease had no cure. In the short term, I could do physical therapy. I made an appointment the next day.

I seemed to be one of the few patients who wasn't there to rehab a new knee or hip. I wore my work clothes—dress shirt and khakis. Jamie Tridico, the physical therapist and owner, guided me through a series of exercises with a heavy ball and elastic bands. It was like a mini-boot-camp. I was sweating. Next time, I would wear a T-shirt and shorts. As I left, Jamie handed me a flyer. It was for a boxing class, twice a week, ninety minutes each session. Boxing seemed counterintuitive. You're going to hit people with a neurological disease upside the head? Wasn't that the cause of Muhammad Ali's Parkinson's? I decided to give it a shot.

The gym was in the worst part of town, in a former warehouse, beside a drug detox center. Protective headgear, battle ropes, and knuckle tape were strewn about. Giant metal fans rumbled against the stifling heat. Dust and grime coated the cement floor. The gym smelled of dried sweat, vomit, and Pine-Sol. A makeshift ring with a plywood base was about four inches off the floor. It did have ropes. And corner buckles. Heavy bags of assorted colors, weights, and quality dangled on thick chains from metal girders that crisscrossed the ceiling. Someone handed me a pair of gloves.

About twenty of us there had Parkinson's, men and women, ages thirty-seven to eighty-five. We lined up at the heavy bags. Jamie, who led the class, barked out the combinations: jab, cross, hook, uppercut; rapid-fire blows, body shots, knockout punches.

"Knock 'em out!" Jamie shouted. "Last round. Knock 'em out!"

An older woman, Ann, swatted at the bag, holding herself upright with a walker between rounds. Volunteers would lift you up if you fell down. After thirty minutes, I was gasping for air. After an hour, my eyes were bloodshot, my arms heavy, my legs wobbly. In my mind, Parkinson's was unrelenting, like a raging fighter coming right at you, but in this gym, you could fight back. Intense physical exercise, short bursts of energy, helped delay the disease. I put my head down, gritted my teeth, and slammed the heavy bag.

17

A Legal Cartel

Sticking my "intrusive journalistic nose" into *State of West Virginia v. AmerisourceBergen Drug Corp., et al.* (as the lawsuit was styled) didn't stop the case from barreling toward its January 2017 trial date. In preparation, lawyers from both sides had a chance to question witnesses expected to testify. During depositions, lawyers could ascertain what witnesses knew, discover new facts, probe for inconsistencies. The exercise would limit surprises at trial. The attorneys would have all their ducks in a row.

One morning in July 2016, Al Emch, lead counsel for the distributors, was grilling Sergeant Mike Smith in a brightly lit conference room at Jackson Kelly PLLC, West Virginia's go-to law firm for coal companies and out-of-state corporations. Its clients included AmerisourceBergen and Miami-Luken. Emch was trying to get Smith, the West Virginia State Police's top drug investigator, to explain why he never picked up the phone and alerted drug distributors about their lawbreaking customers that were buying up millions of pain pills and filling fraudulent prescriptions. Smith said he wouldn't even have known whom to call. The distributors' sales managers made themselves scarce in West Virginia. He never spoke to one.

"No, sir," said Smith. "I never saw any of them. I never met any of them. Which was odd because I figured, you know, as blatant and obvious as the stuff was down there that they would have sent somebody down, like an investigator or something, to see why all these pills are coming to West Virginia. You know, questioned it and saw for themselves what was going on. But, no, I never saw anybody."

This was the distributors' chance to question the state's principal witness before the state's lawsuit against Amerisource-Bergen went to trial in January, to expose weaknesses in Smith's testimony. Jim Cagle, who was still representing the state despite his disputes with the attorney general, had recruited Smith as a witness. Smith knew more about the prescription drug crisis in southern West Virginia than anybody else. He led the bust that shut down the Wellness Center pain clinic in Williamson. He arrested Dr. Donald Kiser. He organized undercover operations for the Highway 119 drug task force—named after the four-lane that stretched through the state's southern coalfields. He was the distributors' worst nightmare: a clean-cut cop straight from central casting, tough, but by the book, who would connect with jurors eager to punish those responsible for the opioid epidemic.

The distributors had Smith under oath. They could ask him anything. Allen Lopus, a Pittsburgh lawyer representing regional distributor TopRx, took over where Emch had left off.

"Sergeant, are you aware of any situation where any of the wholesale distributors provided controlled substances to unlicensed pharmacies or physicians in West Virginia?"

"I don't know anything—I don't know anything about them at all."

"And when you say 'them,' you're referring to the distributors?"

"To the distributors. I guess somebody was telling me they have, like, compliance officers or law enforcement security-type people. I've never met any of those. I wouldn't know anything about them."

The only time Smith had ever heard from a distributor

was when Miami-Luken, with no explanation, started faxing suspicious-drug-order reports to state police headquarters. That was a couple of months ago. He had never before seen one of the reports.

"I'm thinking, jeez, that would have been nice to have ten years ago, before everybody got addicted to prescription pills in West Virginia," said Smith. "I'm thinking this is neat. . . . That's my only contact with the pharmaceutical people or the distributors."

The distributors had two layers of separation between their companies and drug abusers. Their argument was that they wouldn't have shipped so many pills if doctors hadn't prescribed them and pharmacies hadn't placed orders. It was perhaps their best defense. Rebecca Betts, a lawyer for distributor H. D. Smith, wanted to drive that point home, to hear it from the state's star witness.

"You've never identified any individual who received such an illegally diverted prescription drug that was from an order filled by H. D. Smith? You can't name anyone for me today?"

"No. I can't name anybody."

"Do you know if you have the ability to even identify anyone if you were tasked with that today?"

"I don't know how I would do that."

"Thank you."

A lawyer with Masters Pharmaceutical picked up on the same line of questioning: "Have you ever confiscated any controlled substance and traced it back to a wholesale distributor?"

"No." Smith had filed complaints against five pharmacies that ordered massive numbers of pain pills from distributors. He accused the drugstores of filling bogus prescriptions for a sham pain management clinic in Williamson.

"What did the Board of Pharmacy do against those five pharmacies?"

"Nothing."

That's the answer the distributors wanted. State and federal regulators hadn't done their jobs. They should have put a stop to the pill shipments, sounded an alarm, notified the distributors that the orders were getting out of hand. Only the agencies had access to the databases used to monitor the sales. But the pharmacy board did nothing. And the DEA, where was the DEA? How many agents did the DEA have on the ground to stop painkillers from being abused in West Virginia? For many years, just one, Smith told the lawyers, during a decade when pills spread over the state like a hailstorm.

The company lawyers set other trip wires. Emch pounced when Smith mentioned that some West Virginians traveled to other states to buy prescription drugs and then returned to sell them on the black market. West Virginia's prescription drug problem, the oversupply of painkillers, wasn't the distributors' own doing; rather, the companies could blame criminals who shuttled addictive pain pills from the Sunshine State to the Mountain State.

"You were familiar with the Florida connection?" Emch asked. "Flamingo Express?"

Smith dismissed the distributors' notion that the bulk of prescription painkillers saturating southern West Virginia had come from pain clinics and pharmacies in Florida.

"It wasn't too bad because we had our own doctors. I mean, we had our own pain clinics right here. So we did have people who went to Florida. But a lot of them didn't have to because we had our own. We had a smorgasbord of pill doctors down here."

Emch kept inquiring about drug task force investigations. Weren't the addicts at fault? Weren't they the ones abusing the drugs, selling them, breaking the law? It was another defense: Distributors couldn't control criminal behavior.

"In southern West Virginia, the prescription pharmaceuticals—like Lortab, Vicodin, hydrocodone—became its own system of barter," Smith said. "I've had witnesses tell me that if they left the house with a ten-milligram hydrocodone tablet, it was no different than leaving with ten dollars in your wallet. You could exchange it for guns. You could buy groceries with it. It was the same as money."

The lawyers also had questions about how the lawsuits against the drug distributors started. Whose idea was it? Did Smith ever talk to Darrell McGraw? Did he advise the attorney general to file suit? In fact, there was a meeting. McGraw announced he was going after drug companies.

Emch solicited Smith's thoughts. "What was your opinion that you expressed to Mr. McGraw?"

"I told him that I thought pharmaceutical companies were similar to the drug cartels in Mexico."

Emch insisted on a clarification. Surely Smith wasn't including distributors. He couldn't be insinuating Emch's clients were part of some crime syndicate. "And when you say *pharmaceutical company*, what do you mean?"

"The people that are providing the pharmaceuticals or creating the pharmaceuticals or distributing those to pharmacies."

"Wait a minute. Wait a minute." Emch wasn't going to stand for this cartel comparison. Drug cartels weren't licensed. They operated outside the law. They were dope smugglers. They bribed government officials and killed their competition. There was no comparison.

"So what's your state police definition of a cartel?" Emch asked.

"When I'm talking about a cartel, I'm referring to the ones, say, the Medellín with Pablo Escobar, where you have multiple people that are involved in the same conspiracy. And the ultimate goal is to sell drugs and produce as much money as you can. And then you branch out into corruption. And then eventually the cartel becomes so strong that it actually starts to rival the state and federal government. And then the cartel actually becomes part of the government."

Everybody was making money—the pharmacies, doctors,

patients, distributors, manufacturers. And nobody had the power to stop them.

"It's a happy marriage," Smith said. "There's no incentive for anybody to regulate this stuff."

Emch said he wanted to understand Smith's testimony. Was Smith alleging the DEA had been neutered, that it was unwilling to stop drug companies from breaking the law?

Yes, that's exactly what Smith was saying. "I think the pharmaceutical industry has got so much money that they've got the DEA scared. You have so much money coming in with lobbyists and the money that's generated from pharmaceutical sales that nobody wants to take on these people. It's a cartel. They're protected. And you can't—it's too big."

Rod Jackson's black Gucci loafers were annoying the hell out of David Potters, the director of the West Virginia pharmacy board. Jackson had his feet propped up on the conference room table at the law offices of Jackson (no relation) Kelly, where Smith had testified a week earlier. The bald-headed trial lawyer was holding a Cuban cigar, and was he going to light it? No, he was chewing it into a wet mess, spitting the juice into a Styrofoam cup. Jackson—he introduced himself as "P. Rodney Jackson"—had pulled the same stunt when he stopped by the pharmacy board office to chat with its executive director the week before. Potters disliked him immediately. Jackson, who was helping Cagle with the state's lawsuits against the distributors, was obnoxious, uncouth, a show-off, an embarrassment to the legal profession. Potters wasn't fond of Jim Cagle either, but at least Cagle, who had peppered Potters with sharp questions the day before, showed some semblance of decorum. Cagle wasn't wearing blue jeans and masticating a cigar. On this, Potters's second day of testimony, Cagle ceded the stage to his orally fixated cocounsel.

"Let me ask you, Mr. Potters, why are controlled substances controlled?" Jackson asked.

What kind of dumb question was that? This was a deposition, not an exam. Potters knew the answer. One of the pharmacy board's primary functions was to regulate controlled substances.

"Because they're subject to abuse and addiction," Potters said.

"And if they weren't controlled, you could just go to a drugstore and ask for whatever you wanted in regard to a controlled substance?"

"No, they would be prescription only."

"Do you believe that the distributors have a duty to control that dangerous instrumentality that is the controlled substances?"

"Yeah. Their duty is to keep it under lock and key, to keep it protected, to only sell it to other proper registrants and to report the data to the DEA. And then the suspicious-order reports certainly to the DEA. We have the reg in place that we never enforced, but it was there to control that system of distribution."

That was a problem. How could you call out the distributors for ignoring a regulation if the pharmacy board neglected to enforce it? Over two days, Jackson and Cagle asked Potters pointed questions, but his answers spun everywhere. He had an independent streak, and it was on overdrive. Jackson cringed at what might happen if Potters repeated his testimony to the Boone County jury. There was little doubt he would.

For one, Potters didn't believe West Virginia had the highest overdose rate in the United States. Instead, he believed West Virginia was one of the best states at counting overdose deaths: "I know that it is reported that West Virginia has the highest per capita drug overdose rate in the nation. My personal belief is that's in part due to the fact that West Virginia is doing a very good job of investigating, and the chief medical examiner's office is doing the autopsies on those and creating the statistics that maybe some other states aren't."

This was news to the attorneys in the room. The CDC, the state Health Statistics Center, and every overdose study in recent years had West Virginia number one, without a caveat or asterisk that we were doing a better job of tallying the death toll. West Virginia's ranking for economic growth, health outcomes, education, you name it, whatever the issue, was always abysmal compared to other states'. Thank God for Mississippi, we'd always say. But here was Potters claiming that we were one of the best at something, no matter that West Virginia's overdose stats were 33 percent higher than Ohio's, the state with the second-highest rate. It wasn't even close.

"I think we're doing a really good job of investigating those and creating statistics," said Potters, who had served dual roles as the agency's executive director and general counsel for nearly a decade. "Was West Virginia really number two, number three, number four? I'm just suggesting that some other states may not be doing as good investigating all of their deaths to determine if they're drug overdose deaths or not."

To make matters worse for the state's case, Potters didn't dispute a previous witness's testimony that pharmacy board employees had told at least one distributor that it didn't need to submit reports about suspicious orders of painkillers. Distributors claimed they had emails to prove it. Betts, the lawyer for H. D. Smith, asked Potters for an explanation.

"I just learned of this two days ago," he said. "And it is not something that it would seem to me that staff would have or should have said. I can try and check in with retired inspectors, because it could have been a retired inspector or someone who may or may not have said something like that."

Great, Jackson was thinking, these are the same guys who called Sav-Rite and other pill mills "well-managed facilities" in reports to the board.

Then Al Emch had his chance with David Potters, and nobody in the room knew him better. "Dave, from your perspective, given all of your experience, where would you put the bull's-eye—what's the one place, the one action in this system, this closed system, about which the axis of drug diversion of controlled substances circles?"

Potters wanted no part of the state's lawsuit. It was a nuisance, a burden, an assault on his credibility. Records requests had to be fulfilled, depositions given, and his admission that the board neglected to enforce the suspicious-order reporting rules threatened his job. The board was stuck in the middle, and Potters didn't want to take sides.

"My opinion is different than most people's," Potters said. "I put the bull's-eye on the individual that's abusing the drug and seeking it and going out and getting it."

Potters didn't point a finger at the distributors or manufacturers or pharmacies or even the doctors. He was blaming the addicts. That was the last thing Rod Jackson wanted a jury to hear. That argument played into the hands of the distributors. It got them off the hook.

With the jury trial in the state's suit against AmerisourceBergen set to start in six weeks, the distributors had to find someone to rebut Sergeant Mike Smith's testimony. Their lawyers had spent nine hours drilling Smith on their home turf, and he wouldn't give an inch. The distributors hired Mary Rochee, a retired DEA agent with twenty-eight years of service. Between 2007 and 2013—her last assignment before retirement—she led a DEA field office's antidiversion program, with responsibility for curbing the flow of pain pills to dealers and abusers. The field office's territory covered the District of Columbia and three neighboring states, including West Virginia. The distributors were paying her $300 an hour as an expert witness. For years, she was in charge of stopping the drug problem in West Virginia, and now she was working for the companies accused of starting it.

Cagle trekked to downtown Philadelphia in October 2016, to the offices of one of the premier law firms in the city, where he'd have a chance to question Rochee under oath and take her deposition before she took the stand in Boone County in January. He had to figure out a way to disprove Rochee's claims about the causes of the state's prescription drug problem before she testified in front of the jury. His work was cut out for him. AmerisourceBergen couldn't have found a better witness to toe the company line. Before being promoted to a management job at the DEA, Rochee was a frontline investigator in West Virginia. She had worked cases with Mike Smith. She was on a first-name basis with employees at the state medical and pharmacy boards, at the poison control center and medical examiner's office, and at the US Attorney's Office in Charleston. She also had won three awards for exceptional service.

Cagle was going to challenge her assertion that distributors weren't at fault for the flood of drugs into the state, but rather drug traffickers who brought the addictive pills from Florida and West Virginians themselves who traveled to out-of-state pharmacies only to return and sell their stash of painkillers in towns and hollows.

"Just a couple of things," Cagle said at the start. "If I ask you a question, as I often do, I'll lapse into tongues. It's religious, and you may not understand what I say. If you don't understand it, please ask me to repeat it, and I will repeat it, and I will accord you that courtesy, OK?"

Cagle cracked a grin, but nobody laughed. The Philadelphia lawyers in the room didn't know what to make of the hillbilly attorney from West Virginia.

"All right. Thank you," Rochee said. It was her first time as an expert witness, though she had testified about investigations in her capacity as a DEA agent. Weeks earlier, she had put together a three-page report on West Virginia's prescription drug problem, after reading news articles, research papers, and Mike Smith's deposition. AmerisourceBergen shared the report with Cagle. Rochee spotlighted her participation in a confidential DEA operation called the Special Field Intelligence Project in 2010. The project assessed the "who, what, when, how and why" of West Virginia's drug epidemic, documenting "the origin and progression of WV's pharmaceutical drug issue, how it spread into and throughout the state, the various perpetrators and their methods, along with how every aspect of communities were impacted."

The 2010 DEA study, Rochee told Cagle, supported her conclusions in the report she prepared for AmerisourceBergen: Independent drug-trafficking organizations accounted for the most significant volumes of painkillers diverted across West Virginia. Also, some doctors prescribed excessive numbers of painkillers and sold prescriptions for cash and sex. What's more, some people stole from their friends' and families' medicine cabinets, and pharmacy workers pilfered drugs from their employers, and thieves broke into medical offices and drugstores in the dark of night. Delivery trucks were being hijacked, too. Those were the primary causes of West Virginia's prescription drug problem, Rochee insisted.

"Do you remember where these hijackings occurred?" Cagle asked.

"At this particular time, no, I do not."

"Now, what do you remember about the volume of in-transit hijacking, drug shipment thefts?"

"I can't tell you that I have a memory of that."

Cagle asked her for numbers, statistics, anything to back up her report—since she was standing by it—and her conclusion that the pain pills being abused by West Virginians came from other states, hauled here by drug traffickers. Rochee had no numbers.

"Well, let me give you a number," Cagle said. "Rounded, 119 million dosages." That was the number of hydrocodone and oxycodone pills that AmerisourceBergen supplied to West Virginia pharmacies from 2007 to 2012. Did Rochee believe drug dealers were ferrying more pills than that to the state?

"That's beyond the scope of what I'm reporting here."

"Well, I beg to differ with that, but that seems to be what you are reporting here."

"It wasn't what I was researching." She was researching drug traffickers, dealers, abusers, and thieves—not companies licensed and registered to conduct lawful business. "I don't look at our legitimately registered drug wholesalers as drug traffickers. I'm sorry, OK? They're engaged in legitimate commerce, and we don't classify them as drug traffickers."

Cagle wanted to know more about the confidential DEA research. In 2010, the federal agency resolved to get to the bottom of the prescription drug problem in West Virginia, Rochee said. She and another investigator, along with two or three intelligence analysts, were assigned to the top-secret project. The team descended on Charleston. They interviewed state police officers, federal prosecutors, and medical examiners. They spent more than a year collecting evidence and analyzing it. They wrote a report. It was never released to the public. Cagle asked how he could get his hands on it.

Rochee didn't bring a copy to Philadelphia. She wouldn't say if she had one. "I guess you would have to go to the DEA, because I haven't worked there in three years, sir."

"Well, between slim and none, what are my chances of getting that from the DEA?"

"I would say possibly slim, to be honest."

The DEA had shelved a report about an epidemic in the state with the highest drug overdose death rate in the nation, and now a retired agent—someone who was integral to the project—was using its conclusions to defend the companies that stood accused of fueling the public health crisis. Cagle was at a disadvantage to question the "phantom report," as he called it, or Rochee's recollection of its findings, because it remained secreted at the DEA's field office in Washington.

"Do you believe that the DEA has a negative role in the creation of the drug epidemic that we have either here or nationally?" Cagle asked.

Rochee could have said no. She didn't. "I don't have an opinion about that at all."

18

Home Court

Tim Conaway had Judge Thompson's full attention. Now all Tim had to do was persuade him to issue a ruling that would draw the wrath of a $102.5 billion company, Cardinal Health. On this Wednesday morning in early November 2016, Conaway was going it alone in the courtroom—McGinley was in California visiting his grandkids. Conaway was representing the *Gazette-Mail* in its dogged struggle to unseal the state's lawsuit against Cardinal, which remained separate from the AmerisourceBergen case that Thompson had unsealed six months earlier. I had heard whispers that Cardinal was trying to settle with the state to keep its painkiller shipments hidden forever. Conaway figured this might be our only chance to get them. We had to move quickly.

For thirty-five years, Conaway had worked as a personal injury lawyer in Boone County. He represented miners severely injured or killed in mining accidents. His clients included many snared by addiction. Though he had never argued a public records case, he was a quick study. Conaway was going up against a drug distributor with unlimited resources, willing to do whatever it took to win. But Cardinal would have to defeat him on his home court. On paper, he was representing a newspaper. In his mind, he also was advocating for his home county, his friends, his neighbors.

151

He wanted to get to the bottom of what cursed his community. Now, he would have his chance.

"The *Gazette-Mail*'s role is to disseminate this information to the public because undoubtedly this court realizes the seriousness of this drug problem," Conaway said. "I don't think I have to tell the court that. This court was one of the first courts in the state to establish a drug court—and I understand that this court spends a lot of time dealing with this problem." Judge Thompson's criminal caseload was getting bigger by the day, and many of the crimes were drug related. "So I don't think I have to argue to the court how serious this is and how badly it needs to be solved, and the need for data so we can devise intelligent decisions. That's data for the voters, for the scientific community, for the health care community. More information is a good thing."

Conaway took umbrage at Cardinal's assertions in court filings that disclosing the pill numbers would taint the pool of Boone County jurors that would hear the case when it went to trial. Businesses typically want to tout higher sales, Conaway reminded the judge. Ford dealerships brag about how many pickup trucks they sell. "So why aren't these defendants bragging about how many drugs they shipped here? The reason is because they understand that it will be seen by the public as nefarious activity."

Cardinal attorney Henry Jernigan jumped out of his chair. He was ready to explode. He wasn't going to let some bush-league lawyer get away with such an outrageous comparison. Didn't Conaway know that the federal government barred drug distributors from disclosing their sales numbers publicly? "So to suggest we ought to be out here bragging, that's nice. But we clearly are not selling Fords. We're selling a regulated prescription medication that the DEA sets the quotas for each year, sets what the country needs. And certainly there's no allegation that Cardinal or the industry itself has ever exceeded those quotas." He was referring to national production caps that limit the number of painkillers and other controlled substances drugmakers can manufacture each year. The DEA set the number based on past sales and market-trend data. Cardinal distributed drugs. It didn't manufacture them as did, say, Purdue Pharma, makers of OxyContin.

Jernigan warned that making the lawsuit's details public would tarnish the reputations of law-abiding pharmacies that purchased drugs from Cardinal. "And I'm sure that the *Gazette-Mail* in informing the public—should this information be released, and we do not think it will be released—we're sure they will inform the public that these pharmacies, despite the allegations in the complaint, have never had their licenses revoked by the State of West Virginia or the Drug Enforcement Administration."

Jernigan pointed a finger at doctors and licensing boards and regulators. They were the bad guys, not the distributors. Why weren't they being held responsible? The distributors had been making this blame-shifting argument for years. "The State of West Virginia, if they believed that those doctors were acting as pill mills, could have shut those doctors down in a minute," Jernigan said to the judge. "The fact of the matter is, the State of West Virginia didn't do that." And where was the DEA and state Board of Pharmacy? "They never raised any red flags with Cardinal in terms of its shipments. They knew what our shipments were, and yet they did nothing."

Judge Thompson gave Conaway the last word. People deserve to know how many drugs are being shipped to the communities they live in, he said. "It's valuable knowledge." They deserved to know the truth.

"All right." Thompson had heard enough. He was ready to rule.

A month earlier, four days after the *Gazette-Mail* had filed its motion to unseal, Jernigan hatched a deal that he contended made the newspaper's request moot. Cardinal and Morrisey's office quietly agreed to put an indefinite halt to the lawsuit. The state's AmerisourceBergen suit, which had been filed first and always taken precedence over its Cardinal case, was headed to trial in January, and the attorney general's office said it needed more time to prepare. This put the brakes on the Cardinal suit. No more hearings. No more depositions. No more demands for documents or any "related matters," as the court-ordered agreement was worded. Cardinal wasn't going to let some two-bit newspaper and its volunteer lawyers do to it what they did to the other wholesale distributors. Its pill shipments to individual pharmacies would remain confidential—never mind that its peers had lost that same battle.

We were unaware of the deal to put the Cardinal case on hold until Jernigan informed us in an email to the court. After Morrisey had taken office in 2013, Cardinal abruptly fired its Charleston-based defense team and hired Jernigan, a $500 donor to Morrisey's inauguration party.

"It was our understanding, based upon the recently entered order, that proceedings were stayed pending the disposition of the AmerisourceBergen case," wrote Jernigan in the email. "We would suggest the pending motion be deferred consistent with that order."

The court order had stopped "all discovery (depositions, document requests) and related matters," and Jernigan argued that the *Gazette-Mail*'s request to step into the state's lawsuit against Cardinal was a related matter. Nothing had been filed in the docket for months. The state wasn't even responding to Cardinal's requests to turn over records. The case was on unofficial hiatus, Jernigan claimed. The newspaper was trying to undermine the agreement. It would have to wait until after the AmerisourceBergen trial, and then the *Gazette-Mail* could make all the motions it wanted.

The court-ordered pause may have been enough to reject the newspaper's request to intervene, but Jernigan had more reasons: Cardinal should be given the chance to settle the lawsuit and keep its pill numbers cloaked—the same opportunity the judge had afforded other wholesale distributors earlier in the year. Jernigan reiterated that Cardinal delivered pills only to state-licensed pharmacies that filled prescriptions from state-licensed doctors. A hearing wasn't needed, the court had no reason to unseal the revised lawsuit and its new allegations, as far as Cardinal was concerned. Nothing good would come of it.

"The impression created by the publication of these allegations by the *Gazette-Mail* or any media will undoubtedly be that Cardinal Health was unlawfully shipping thousands and thousands of prescription medications to pharmacies that were criminal or otherwise illegitimate," Jernigan argued.

Judge Thompson scheduled the hearing anyway.

After finishing their closing statements that day in early November, Jernigan and Conaway sat down in the hushed courtroom. As the judge began to speak, I leaned forward in my seat, poised to take notes, straining to hear him in the cavernous courtroom. Conaway was tapping his pen on a legal pad.

Thompson announced he was dismissing Cardinal's argument that the three-week-old agreement with the attorney general halted all proceedings. He rejected Cardinal's request for time to settle the lawsuit, given that the two sides hadn't even started settlement talks. He conceded that Jernigan had made a good argument about unfairly tainting pharmacies, but "not necessarily a winning argument." As for the jury pool, the allegations in any lawsuit are one-sided. It's up to the defense team to rebut them. "This is not two private corporations suing one another, or an individual suing another individual or corporation," Thompson said. "This is a state action filed on behalf of the citizens of West Virginia. I think it should be more transparent and open."

He directed Conaway to prepare a proposed court order. The judge was going to unseal the court filing. Two days later, he did.

I didn't find out that Thompson had issued his order to unseal until three o'clock on a Friday afternoon. The Boone County

Courthouse was closing in an hour. I tried to call Conaway, but was unable to reach him. So I jumped in my Honda Civic at the capitol and gunned it along the highway to Madison. I dashed to the clerk's window with two minutes to spare before closing time. The clerk made a copy of the unsealed lawsuit and handed it to me. The thirty-one pages felt heavier than they actually were, perhaps because of the struggle, against long odds, to get it. I sat in my car, flipping through the pages. Nothing was blacked out, nothing was redacted. Cardinal was the top seller of prescription painkillers in West Virginia. It had saturated the state with hydrocodone and oxycodone—a combined 240 million pills between 2007 and 2012. That amounted to 130 pain pills for every resident. The lawsuit cited huge shipments to the counties most affected by the drug problem: Logan, McDowell, Boone, and Mingo. These counties were in the heart of the state's southern coalfields. The region was shedding population, towns were dying, people were overdosing in record numbers. Hope was hard to find. In Mingo, Cardinal delivered 2.1 million hydrocodone pills to Williamson over six years, and 207,000 pills over two years to Van, a Boone County town with 241 people. Other large shipments went to two troubled drugstores whose pharmacists later served prison time for filling illegal prescriptions.

After driving back to the capitol, I picked up the phone in the pressroom and called Cardinal's corporate spokeswoman for comment. Near the end of our conversation, she politely asked me to correct an error I had repeated in previous stories. I kept referring to Morrisey's wife as a Cardinal lobbyist. That was wrong. The company had severed ties with Denise Henry Morrisey five months earlier, the spokeswoman said. That, I realized later, coincided with a CBS News investigative report about Morrisey's ties to Cardinal Health that aired the same week.

During an interview with Morrisey at his office in the capitol, CBS reporter Jim Axelrod pressed the attorney general with sharp questions.

"While you've been in office, your wife's firm has made roughly a million and a half bucks from Cardinal," Axelrod said.

"Ah, but—" Morrisey stammered, his lips quivering. "You'd have to talk and take a look at those numbers. I don't pay attention."

"So how does that not present an enormous appearance of a conflict, if not an actual conflict?"

"Well, I think we've gone through this, the process, and people have determined there was no conflict." Morrisey smiled.

Axelrod closed his report by reading from the West Virginia lawyer disciplinary board's findings, which concluded Morrisey's wife's lobbying for Cardinal while his office was suing the company "could diminish the integrity of the process and create the appearance of impropriety." It was one thing for a small newspaper in West Virginia to be publishing stories about Morrisey and Cardinal, but another for CBS to be broadcasting the story to the world in prime time.

Cardinal's spokeswoman wouldn't tell me why it had terminated the contract with Denise Henry Morrisey and instead emailed a company statement: "She did not represent Cardinal Health on prescription drug legislation at the federal or state level." We printed the statement verbatim in the story. I made a mental note to check on it later.

19

High Noon

It wasn't a question Patrick Morrisey was expecting at the open-air debate he had orchestrated for his reelection campaign.

Two weeks before the November 2016 election, Morrisey had taken to Twitter and challenged his Democratic opponent, Doug Reynolds, to meet him at "high noon" on the front steps of the capitol. The media was calling it the "Battle at the Capitol." Nobody was sure whether Reynolds would show. If he didn't, Morrisey would call him a coward. If he did, he might be ambushed. Morrisey's campaign was paying for the production.

Reynolds arrived five minutes early. Tall and athletic, with short blond hair, the forty-year-old Huntington businessman was no stranger at the capitol. He had served ten years in the state House of Delegates. Previously he worked as a county prosecutor. Three microphone stands—the third was presumably for a moderator—and two metal chairs were set atop a landing halfway down the limestone steps. Reynolds stood alone, smiling. The sun flared in the sky. It was unusually warm for the third week of October. Both candidates had supporters in attendance. Morrisey's held up campaign signs. But there was no sign of the attorney general. He had set the trap. Did he think Reynolds wouldn't take the bait?

Just then, Morrisey pushed through the capitol's heavy brass doors and paused at the top of the steps, surveying the crowd. His campaign manager seemed to be coaxing him to get a move on. It was past noon. He hiked down the steps and stopped at one of the microphones. I was sitting on the rim of a fountain, watching the spectacle. The candidates' campaign chiefs huddled, arguing over ground rules.

"We're going to be happy to take questions," Morrisey started. "I appreciate this opportunity to be here today to talk about the issues that face West Virginia."

I did have a question—for later. Morrisey, for ten weeks, had ducked my public records request for the DEA documents referenced in the now-unsealed lawsuit, showing the quantities of pain pills being shipped to West Virginia towns and cities. Morrisey's office kept putting me off. Aides claimed they were searching for the records. They just needed more time. I asked for proof they had started a search. They had none. It was one delay after another. I figured Morrisey would stall until after the election, then all of a sudden find the documents but refuse to release them. He had already defeated us in court once, persuading a judge to keep records about his involvement in the Cardinal lawsuit hidden. He could defeat us again. Nonetheless, Pat McGinley was already preparing a lawsuit to pry the DEA documents loose. Rob Byers, our executive editor, suggested writing a story about Morrisey's refusal to release them. I realized this might be our only chance to press Morrisey. I had passed along the question to our political reporter, Andy Brown, who was waiting with other reporters for Morrisey to finish his opening statement. There was no moderator.

"We created the first substance abuse fighting unit within the office of attorney general," Morrisey said, ". . . positioning the state to be a leader in taking on substance abuse."

Conservative newspapers in the state had praised Morrisey as West Virginia's top opioid fighter. He had sued McKesson, the nation's largest drug distributor, ten months earlier after two years of delays. He had also begun promoting select high school football matchups as "Opioid Abuse Prevention Games of the Week," assigning his employees to information booths at the stadiums. They would hand out materials on the dangers of prescription painkillers. The program was advertised on radio stations across the state. He was trying to turn his political Achilles' heel into an asset.

"Our best-practices initiative is going to dramatically reduce opiate abuse by twenty-five to forty percent," Morrisey said, his opening statement dragging on, "and that's why it's earned the support of twenty-five state and national groups." What had his opponent done to tackle the crisis? Morrisey couldn't resist telling. He jabbed a finger at him. "Doug Reynolds also has no record to speak of fighting substance abuse. While Huntington is the epicenter of the substance abuse problem in our state, Doug Reynolds, as a legislator, did not support one piece of legislation as a sponsor to fix the problem. He was too busy trying to garner more government contracts for himself and his family."

Reynolds's father owned a printing company that did business with state agencies, and Reynolds owned the newspaper in Huntington, and he was CEO of a pipeline construction company. He hadn't practiced law in years. He had provoked Morrisey's wrath after airing campaign ads about the attorney general's wife living in Virginia and lobbying for the pharmaceutical industry. Reynolds's TV ads also included clips from the unflattering CBS News report about Morrisey's wife's lobbying for Cardinal Health. Morrisey called his Democratic opponent "Deceitful Doug Reynolds."

"This is a man that cannot be trusted," Morrisey said, "and this is a very important job. That's why I'm eager to have a debate on the issues and to learn more, finally, about where Doug Reynolds stands."

Reynolds had stood there silent, listening to Morrisey's diatribe. When he finally got a chance to speak, Reynolds brought up his job for five years as a prosecutor. Morrisey had coveted the power to prosecute since the day he took office. He wanted to be a crime fighter, but the state constitution wouldn't allow it. "I'm the only candidate that's worked in the trenches as a prosecuting attorney and actually put drug dealers in jail," Reynolds said, his voice getting louder. There were some cheers, a smattering of applause. Morrisey stared at him, fingers laced. Reynolds had helped start a drug abuse treatment center in Huntington. He defended his record as a state legislator. "Now, the attorney general talked about substance abuse and what his office has done. The last four years we have lost the battle with drugs. Can anyone really deny that? We look down in southern West Virginia, what's going on down there, 241 million doses." Reynolds was citing Cardinal Health's painkiller sales numbers from the court documents we had successfully unsealed.

During the time for questions and answers, the two went back and forth over who was a bigger champion of the coal industry and would fight harder against the Environmental Protection Agency. Morrisey touted his legal battles against transgender bathroom laws and immigration. He criticized Reynolds for trying to push through an ethics bill that would have reined in Morrisey's ability to file lawsuits without the governor's permission. Reynolds accused Morrisey of spending $150,000 in state funds to advertise his opioid football games of the week.

A woman in the crowd asked about Reynolds's donating to Hillary Clinton's campaign. He gave Clinton $2,300 in 2007 while she was running against Barack Obama in the Democratic presidential primary. "I have not donated to Hillary Clinton since Mr. Morrisey has been a practicing lawyer in the state of West Virginia," Reynolds said.

"Let's make sure we know the record here," Morrisey jumped in. "Doug Reynolds donated twenty-three hundred dollars to Hillary Clinton. Doug Reynolds also donated to Barack Obama. That's Doug Reynolds's record, and he can't run away from it."

The Republican Attorneys General Association (RAGA) had

pumped more than $6 million into the West Virginia contest, mostly on television advertising that linked Reynolds to Clinton and Obama. Both were wildly unpopular in West Virginia. The commercials ran nonstop. Reynolds donated $4 million of his own money to his campaign, but he couldn't keep up with RAGA's spending spree. Who were RAGA's donors? They included Purdue Pharma, McKesson, and Cardinal Health. The organization also paid a tracker to follow Reynolds.

"Starting in March, there was a gentleman standing right there." Reynolds pointed at him. "He's filming right in front of me now, started following me and my family around." The man with the camera waved. "Who is this guy? He's paid for by the pharmaceutical industry. They've shown up at parks, fairs, and festivals, harassed my three-year-old daughter. So let's start with how it started."

Morrisey called on Reynolds to apologize for bringing Morrisey's wife into the campaign and said, "I'm not going to bring your wife and kid into the race."

"Stop lying, Doug," a man shouted.

"Wait a second," Reynolds said. "Your wife is a lobbyist for Cardinal Health."

"He doesn't address the question," said Morrisey. "Note that."

"No, no, no, no," Reynolds shot back. "While he was attorney general, she and her firm made one and a half million dollars in the case that he's supposed to be protecting us from these people."

"When I took office, I did the right thing," Morrisey said. "I stepped away from cases. We handled things right. Do you know my wife used to represent companies for many years long before I was the attorney general having nothing to do with the substance abuse epidemic in West Virginia?"

Reynolds turned to the audience, saying, "Can we ask a question about drug addiction?"

Andy, a pit bull of a reporter whom I'd helped recruit to the *Gazette-Mail*, was ready with the question that Morrisey wouldn't have prepared an answer for. I stood up and stepped closer to listen. Here was our chance.

"The question is, from a societal standpoint, to understand the crisis, you have to have information on it. Why is it that you're not releasing DEA records that were cited in a lawsuit, and would you push in the future to make sure that all of the information is brought forward and that the public knows about it, not just the lawyers who are litigating the cases?"

"Yet another lie from my opponent," Morrisey said. There was booing. "Do you know I have always stood for transparency?" In a rambling answer, Morrisey blamed Judge Thompson for blocking Morrisey from giving us the DEA data. The judge decided the records had to be kept confidential, he said. "So we have to follow the judge's orders here."

I didn't understand what Morrisey was talking about. Thompson had unsealed the filing that cited the DEA numbers. As far as we knew, Reynolds wasn't even aware that the *Gazette-Mail* had filed a public records request.

"We filed a Freedom of Information Act request for the DEA records that were cited in the cases with your office, and that hasn't been fulfilled," Andy said.

"So we get a lot of FOIAs from your office, as you well know." Morrisey cracked a wry smile. "The *Charleston Gazette* spends a lot of time trying to attack my office. But we respond in kind, and we've always been responsive to this and will continue to be so."

There were additional questions about Morrisey's support for presidential candidate Donald Trump, about climate change and gun rights. I wasn't paying much attention anymore. Andy had asked his question. Morrisey answered it. I had a quote if we decided to go ahead with a story about the attorney general's reluctance to release the DEA records his office was citing in the now-unsealed lawsuits against the drug distributors, records that would presumably shed light on the causes of the opioid crisis— that's what they were calling it now, a crisis, an epidemic. And

I had Morrisey's on-the-record explanation for denying us the records if McGinley received the green light from the *Gazette-Mail*'s owners, the Chilton family, to sue Morrisey.

We did neither. The day after the debate at high noon, Morrisey's office, without explanation, put a large package in the mail, addressed to me. Its contents arrived in my newsroom mailbox two months after my formal request, even though state public records law required a response within five business days. It was the pill numbers the DEA had provided to the attorney general, hundreds of pages that would change everything.

The pages disclosed the number of hydrocodone and oxycodone pills sold to every pharmacy in the state, as well as the drug distributors' shipments to all fifty-five counties in West Virginia. Graphs traced the number of pain pills by strength in milligrams over time. Charts sorted and ranked counties' pill numbers per person. Shipments were broken down by distributor warehouse locations. Tens of thousands of numbers were here.

And also emails. Kyle Wright, an agent at DEA headquarters in Washington, had sent a batch—with pill numbers attached—in early January 2015 to Betty Pullin, an assistant attorney general. Pullin was assigned to the lawsuits against AmerisourceBergen, Cardinal Health, and the other wholesale distributors. She had called the DEA and requested the shipment records to bolster the state's case. Pullin and Jim Cagle both wanted more ammunition. They had a deadline to file new allegations.

"We got it to you under the wire," Wright wrote to Pullin. "Because the data was so voluminous, we broke it down by years."

"Thank you so much!!" Pullin responded. "All of this information is very, very helpful."

It was an understatement. These weren't aggregate painkiller sales numbers. They were granular. You could trace the number of highly addictive pain pills shipped to every single pharmacy in West Virginia, to every county, city, and small town. You could see which distributors, by name, flooded the state's most rural and poorest counties. The numbers were indisputable. Here was an exhaustive, astonishing accounting of the opioid epidemic in West Virginia. I was holding the evidence, the first reporter in America ever to get his hands on such sales data. I couldn't wait to get started.

Then my phone rang, a single ring—meaning a call from the *Gazette-Mail* newsroom, rather than from an outside line. I was scouring DEA records at the capitol pressroom, and an editor was calling with an assignment. I had drawn the short stick to cover a press conference downtown. A Republican state senator had sent a cease-and-desist letter to Charleston-area radio stations, demanding that they stop airing a political attack ad that falsely insinuated the senator had testified on behalf of a convicted child molester. This same senator had won election four years earlier after I wrote a story about his opponent's being a former pro wrestler who posted lewd comments on a message board. I put down the DEA documents and drove to the press conference.

As the *Gazette-Mail*'s statehouse reporter, I covered and wrote daily stories about state government, the legislature, and, yes, testy elections. I wrote about 250 articles a year. Finding time for bigger stories, such as an investigation into the sales practices of America's largest opioid distributors, proved difficult. I envied most reporters at larger newspapers. They got to spend the bulk of their time on in-depth pieces with few daily assignments to distract them. But at a small newspaper such as ours, we had to learn to juggle big stories with the little ones. Our subscribers were counting on fresh local news every day. Our editors demanded it. And we delivered it.

Not until hours later, after I filed the breaking story—the senator called the radio ads a "character assassination" as his wife wiped away tears—did I get a second look at the DEA records. I quickly made an extra copy and hid it in a file cabinet for safekeeping. I had something that mattered in my possession—disclosures that the distributors and pharmacies could not refute. They had reported these numbers. This was their sales data. There was no getting around them. At any moment, I half expected DEA agents to storm the pressroom with a subpoena and take them back.

I leaned over my desk and started paging through the packet. The documents listed the name of each pharmacy, its DEA registration number, its location by town or city and county, and the number of hydrocodone and oxycodone pills shipped and dispensed by year. Those numbers were in the last column. I reviewed the oxycodone numbers first, then switched to the sheets that disclosed hydrocodone sales. Most pharmacies averaged fifty thousand pills or so each year. Wright had sorted the information by county in alphabetical order. I skipped to the counties that started with *M*—Marion, Mason, McDowell, Mercer, Mineral, Mingo. I traced my finger along the column with the pill numbers and stopped halfway down the page. There was a seven-figure number. The print was minuscule, so I placed a ruler on the paper to make sure I wasn't mistaken. I checked the same column for the following year—another seven-digit number. All hydrocodone pills. Both numbers, which hadn't been disclosed in the state's lawsuit against distributors, lined up with the lone pharmacy in Kermit.

I hesitated to tell the newsroom I was sitting on data that could shake the drug industry to its core. The distributors were sending some of their largest shipments to some of the smallest towns in the state. I worried the editors would want the story for the Sunday edition. They agreed to give me time to do more reporting. I had three weeks to put together the story of a lifetime.

20

780 Million Pills, 1,728 Deaths

At a small newspaper, I couldn't count on my article reaching hordes of readers, but I was determined to write something that would make a difference, something that lasted. The story, I knew, was about more than just numbers. To tell it right, I needed to talk to the people most deeply affected: victims and survivors.

Chelsea Carter helped me find them.

In late November 2016, Chelsea was working as a drug counselor at an addiction treatment center in Logan, where I interviewed her about her own demons, about how OxyContin had hijacked her brain a decade earlier. As a young girl, Chelsea had everything. Her father was the mayor of Madison. She was smart and pretty. She was a high school cheerleader. But Chelsea also had a bit of a wild side. Her friends persuaded her to try marijuana. Eventually she began smoking nearly every day. Around that time, Boone County was being saturated with pain pills. You could buy them on the street. You could buy them at the drive-through pharmacy. OxyContin was easy to come by. It became her new best friend. It took away pain, stole her emotions. Chelsea's boyfriend—he was nearly twice her age—supplied her with a steady diet of pills. Soon, she was crushing and snorting them. The high was always better that way. Then she was putting the tablets on a spoon, heating it with a cigarette lighter, melting the pills into a liquid. It wasn't difficult to find a vein. She would pass out and wake up with the needle still stuck in her arm. It was all over then. She was hooked.

She started stealing money from her family to buy more pills. She stole from her friends and neighbors. She was using more and more. Nobody wanted to report her to the police. When you're mayor, you don't want to go to the cops and tell them your daughter is a thief. Her addiction turned to desperation. She broke into cars and sheds and garages. She was arrested and jailed. It wasn't so bad. She learned how to braid her hair in cornrows. For a time, she enrolled in Judge Will Thompson's drug court. It was a pain in the ass. Drug court was work. You had to appear before the judge every month, you had to pee in a cup, you had to find a job. Why bother?

At night, she scoured the streets for trouble. She stole a gun. The charges were more serious. She went back to jail, but this time she dropped to her knees and prayed. "Lord, if you ever bring me out of this, I'll never touch another drug again," she whispered to herself. This time, something was different about the clank of a cell door slamming shut behind her. She'd had enough.

One autumn morning in 2008, she stood in Judge Thompson's courtroom. She wore an orange jumpsuit and handcuffs. Thompson was prepared to sentence her. She faced two to twenty years. Chelsea asked the judge to send her back to drug court. He did.

From that day forward, she was going to do everything right. She was a star pupil in Thompson's drug court the second go-round. She graduated from college and then secured a master's degree. Her felony conviction limited her job options. She got hired as a counselor at the treatment clinic in Logan, where they handed out strips of buprenorphine to wean addicts off heroin and painkillers. Addicts talked to Chelsea. She was one of them. She gave them hope.

• • •

Chelsea suggested I visit her friend Erika Mullins, in Madison. Erika's brother had died of an overdose three years earlier. I coaxed one of our photographers, Sam Owens, to come along. The mid-December deadline to publish the story was closing fast. That, and my Parkinson's diagnosis, focused my attention.

A police cruiser was parked outside the Mullinses' house. Erika's husband, Nick, was a local police officer. They lived in a development of ranch homes. Nick was still asleep after pulling a midnight shift. Their daughter had scattered plastic toys across the carpet and couches and under an artificial Christmas tree. Erika's mother-in-law called to say she was bringing over chili-covered hot dogs.

Erika told us about her brother, Travis. He was an honor student in high school, a baseball player, and Erika thought he would make a good lawyer someday. After graduation, he got a job at Dairy Queen and started running with the wrong crowd, and one day he traded his truck for pills. He'd take Xanax, Opana, and generic oxycodone. One overdose left him in a coma. After he came out, another killed him. He died at his home. He was thirty-one.

"I don't think if he knew how heartbroken we all was he ever would have done drugs," Erika said. "He loved his family, but he had demons he couldn't overcome."

Kay Mullins helped herself into the house, toting lunch wrapped in aluminum foil. Nick shuffled into the living room, rubbing his eyes, squinting from the change of light. We had been talking about one overdose death in the family. Now, I learned of a second.

Kay began reminiscing about her daughter, Mary Kathryn, whom everybody called by her nickname, Goofy. Her addiction began after a car crash a decade earlier, not long after her fortieth birthday. Her back was hurting. A doctor prescribed OxyContin.

"She got messed up," Kay told us, tearing up. Sam put down her camera and handed Kay a tissue. "They wrote her pain pills, and she just got hooked."

Kay had a difficult time recounting the ten years that followed—the lies her daughter told her to cover her addiction, the stealing from her brother, the time she shot herself in the stomach in an attempt to end her life. She'd get 90 to 120 pills and finish them off in a week. She would go to dozens of doctors for prescriptions and stopped at pharmacies in Madison, Logan, and Williamson. She could always find a way to get pills. She kept most for herself, but sold some to others.

"It tore my family up," Kay said. "You don't sleep. One time she would be OK, and you'd think she would come out of it, but then something else happens."

Kay's daughter's hunt for pain pills ended two days before Christmas in 2015. A doctor had prescribed her OxyContin and an antianxiety drug. Two days later, Mary Kathryn stopped breathing in her bed. Nick responded to the 911 call. He tried chest compressions, but he could not revive her. She was fifty. Mary Kathryn left behind a daughter who was struggling with a heroin addiction.

After the funeral, Kay had one last thing to do. She found an appointment reminder card for her daughter's scheduled visit with the doctor who wrote her last prescription. Kay dialed the number. A receptionist answered.

"I told her my daughter was there December twentieth," Kay recalled. "I said, 'Y'all wrote these prescriptions, and she's gone on December twenty-third. I just wanted to let you know she won't be back.'"

The story was missing another dimension: Why hadn't anyone done something to stop the massive flow of deadly pain pills? The West Virginia Board of Pharmacy and the DEA were the gatekeepers. They were charged with flagging pharmacies that ordered a suspicious number of painkillers. Wholesale distributors and pharmacies had to register with the board, and it inspected pharmacies once every two years. Could it have sounded the alarm?

The pharmacy board's office was inside a historic house several blocks east of the capitol in Charleston's East End. An administrative assistant hauled two Bankers Boxes full of faxes into the board's conference room and set them atop a long wooden table. The documents would have alerted regulators to the prescription drug menace years ago—if anybody had bothered to pay attention. I started counting them by hand.

A retired pharmacist had tipped me off to a state rule that required drug distributors to submit reports to the pharmacy board about drugstores that ordered a suspicious number of pain pills. The DEA had the same regulation. I put in a request with the board to see the reports. On this morning in December, they were nestled into the boxes before me. The board's forty-four-year-old executive director, David Potters, stepped into the room. He wasn't there to stop me.

"Do you have some kind of count of these?" I asked. I had known Potters for nearly four years while covering health-related legislation at the capitol. He had always been helpful. He had never tried to hide anything.

"No."

"What do you do with them?"

"We file them away in a drawer. It's a large drawer."

Nobody had ever counted the reports. Nobody at the board had investigated a single suspicious order. Nobody had contacted any drug distributor or pharmacy about a report. The board had done nothing. I asked Potters about the rule, a three-sentence regulation designed to keep in check the flow of prescription pills into the state. Potters ambled over to a bookcase, pulled out a small rule book, and flipped to the regulation in question. It spelled out what orders should be flagged: "those of unusual size, orders deviating substantially from a normal pattern and orders of unusual frequency." The rule had the force and effect of state law.

Potters knew what I was going to ask next. "It wasn't on our radar. It's not ever been an item that's ever been enforced by the board. Nobody brought it to my attention."

Before June 2012, the pharmacy board had received just two reports, Potters told me. But after Darrell McGraw filed the lawsuits against Cardinal, AmerisourceBergen, and other distributors, the suspicious-order reports started arriving in bunches. They were always faxed. The office assistants weren't sure what to do with them. There were so many. Cardinal was faxing about forty a month. McKesson waited until after Morrisey started his investigation. The rule didn't specify what the pharmacy board was supposed to do with the reports. So the board shelved them—every one.

I counted them by hand. Cardinal had submitted at least 2,428, though nine months' worth was missing.

Potters shrugged. Someone likely threw them away. "They were apparently never filed and lost."

By my count, McKesson had identified 4,814 suspicious orders from pharmacies across the state. It faxed over spreadsheets. Masters Pharmaceutical turned in ten, and Smith Drug Company, a regional distributor unaffiliated with H. D. Smith, filed one. Having received more than seventy-two hundred reports in all, the pharmacy board hadn't acted on any of them.

The DEA also requires drug distributors to report suspicious orders. The West Virginia rule was copied almost word for word from the DEA's rule. Potters was quick to point out that the regulation, which had been on the books for almost two decades, didn't specifically name wholesale distributors. It referred to "registrants." The DEA registers wholesale distributors and pharmacies. The pharmacy board also licenses both.

"I think the rule was poorly written," Potters said. "It should have said 'wholesaler.'"

The following week, the pharmacy board held a meeting and voted unanimously to send letters to drug distributors, asking them to report suspicious orders. The board talked about creating a standardized form that the companies could fill out.

"We need to work on this," the board's president said. "We're going to work on this hard."

"For many years, the board didn't really want suspicious-order reports," said a lawyer for wholesale distributor H. D. Smith, shifting blame to the regulator.

A representative of Morrisey's office also addressed the matter with board members. Vaughn Sizemore, a deputy attorney general who was managing the lawsuits against Cardinal, McKesson, and AmerisourceBergen, suggested the board change their rules and require distributors to send their suspicious-order reports to a new location—directly to the attorney general's office.

The deadline was now four days away. I ducked into the newsroom to pick up mail, chat with editors, and select photos for the story I had been piecing together. It would expose how prescription opioids flooded into small-town pharmacies, leaving addiction in their wake.

Andy Brown peeked over the edge of his cubicle and waved me over. "Hey, take a look at this."

Andy, in his spare time, had created two purple-shaded choropleth maps of West Virginia. One map showed the number of painkillers shipped to each county per person. The other represented the thousands of overdose deaths caused by prescription opioids in those counties. The two maps were nearly identical. Counties flooded with pain pills had the highest overdose death rates. The counties shaded the darkest were in southern coalfields.

After the DEA pill records were converted to electronic spreadsheets, they could be sorted, added, ranked, and tracked. On my computer, a six-year-old Acer laptop, I combed through tens of thousands of rows of data late into the night at the capitol. As my Parkinson's progressed, typing and maneuvering the mouse had become increasingly difficult. The tremor in my right hand had worsened, especially when it hovered over the keyboard, slowing my efforts to finish the article. My doctor had prescribed several Parkinson's drugs, but none had proved effective in combating my physical struggles. While reviewing the spreadsheets, I discovered two pharmacies in the same town, just blocks apart, but one was selling ten times as much oxycodone as its competitor. The mom-and-pop pharmacies were the worst offenders—and they all were in southern West Virginia. We had six years of data, so we could rank pharmacies by the number of pills purchased. A Raleigh County pharmacy was shipped the most oxycodone—4.8 million pills. The statewide hydrocodone numbers were even higher. Among individual drugstores, Sav-Rite Pharmacy's 2007 and 2008 purchases stood out. In those two years, the distributors had supplied Jim Wooley's business with nearly 9 million hydrocodone pills—along with tens of thousands of oxycodone pills. That amounted to more than eleven thousand painkillers a year for every resident of Kermit. Other small-town pharmacies were large buyers: Family Discount Pharmacy in Mount Gay, Tug Valley Pharmacy and Hurley Drug in Williamson, and Larry's Drive-In Pharmacy in Madison.

For the first time, we could calculate the six-year total for the entire state. There were two ways to do it. Simply add up all the pills the pharmacies bought. Or add up each distributor's deliveries to every county. Both methods produced the same number—780 million. That was just hydrocodone and oxycodone—in a state with fewer than 1.8 million people. During those same years, 1,728 West Virginians fatally overdosed on those two painkillers, according to data sent to me by epidemiologists at the state Health Statistics Center.

I wasn't sure how to start my story. Not until Andy showed me the choropleth maps and their ominous shades of deep purple, like bruises, did I settle on a first sentence. It came to me at home in the middle of the night. I reached over the side of the bed, fumbled around for a pen, and scribbled on a scrap of paper: *Follow the pills and you'll find the overdose deaths.* I never bothered to turn on the light.

21

Misery's Price Tag

Rod Jackson wasn't going to leave the Boone County Courthouse without a deal. The new year was fast approaching, then jury selection, and a verdict against the distributors wasn't a slam dunk. A billionaire businessman, Jim Justice, had been elected governor, and the makeup of the state Supreme Court had shifted further to the right. There were too many variables. Judge Thompson had ordered mediation sessions before, with nothing to show for it. Neither side would compromise.

This morning was turning out to be no different. Settlement talks were going nowhere, the lawyers were getting grumpy, and Jackson, who was helping Cagle negotiate for the state, wasn't going to get his deal, even though Thompson had taken steps to establish a convivial atmosphere. The judge had sent out for lunch, a catered spread atop a giant table in the courtroom, but it wouldn't change Jackson's sour mood. It was Mexican, and, well, Mexican gave Jackson indigestion.

The mediator, Don O'Dell, had separated the opposing parties into two camps. Jackson, Cagle, and a dozen state lawyers squeezed into a side room normally reserved for jury deliberations. Cardinal Health and AmerisourceBergen lawyers—they agreed to negotiate with the state together, even though their cases were separated—crowded into an office once used by the

177

prosecuting attorney. Both spaces were standing room only. The state's lawyers opened at $100 million. O'Dell delivered the proposal to the distributors' attorneys, who discussed it, dismissed it, and counteroffered $4 million. So it went all morning, O'Dell traipsing back and forth between the two rooms, the floorboards creaking, wooden doors clicking open, heads shaking, doors slamming shut. By noon, the Mexican food had arrived, and the two sides were still $80 million apart.

Jackson was chewing his cigars harder than usual. It helped him relieve stress, and this morning was particularly stressful. He was frustrated. How could they still be so damn far apart? But then, as everyone else was scooping up greasy burritos, salsa, and Spanish rice, an idea hit him. He sidled up to Amerisource-Bergen's general counsel, Elizabeth Campbell, and asked if they could talk in private. They knew each other from previous mediations. Jackson had met her only a couple of times, but he already considered her a friend. After sloughing off another lawyer, Jackson and Campbell were alone in the office.

Jackson got straight to the point. He kept his feet planted firmly on the floor. "Elizabeth, can you give me some idea, a ballpark? Or we're going to be here all day."

Campbell didn't give Jackson a ballpark. She gave him an exact amount. The companies would settle for a combined $36 million. Not a penny more. They were prepared to appeal all the way to the US Supreme Court if they had to. Jackson was astonished. He had a precise number, a big number. He hurried back to the jury room to share the news. It sounded good to everyone. The governor's lawyer, Peter Markham, had to call Tomblin to get his OK. Tomblin wanted $40 million. Jackson took the request back to Campbell. She shook her head. It was $36 million. Final offer. Take it or leave it. They took it. The lawyers had O'Dell draw up a memo. Everyone signed it and shook hands. They could draw up the official paperwork later. Then they'd have to schedule a press conference. They agreed to keep the deal quiet until then.

Judge Thompson stepped out of his chambers. He and Jackson had their picture taken together. Thompson then showed Jackson a sheet of paper with a prediction. Before the mediation began, the judge had jotted down a dollar figure, his best guess at the final settlement. He missed by $1 million.

By then, it was almost three o'clock, and Jackson shot out the courthouse door. The beaches of sunny South Carolina beckoned. Wife number four was waiting at their oceanside villa. But first, he had to head back to his apartment in Charleston to pack. It was a forty-minute drive for most, but thirty minutes in a speeding black Mercedes.

In the second week of January 2017, outgoing Governor Earl Ray Tomblin summoned me to his office to share the news. There would be no trial. He was announcing a settlement with AmerisourceBergen and Cardinal Health. It was reason to proclaim victory—$36 million was the largest pharmaceutical settlement in state history—but something had been bothering the governor. He was leaving office in days, after finishing out a second and final term. He wouldn't be around to ensure the settlement funds were spent on law enforcement and drug treatment programs that helped West Virginians addicted to opioids. He feared the state legislature might try to grab the money and use it to backfill a looming budget deficit.

At Tomblin's request, the settlement papers directed the state's health and public safety agencies to put their portion of the funds into a special account at the state auditor's office. Days earlier, Tomblin had talked with Morrisey, asking him to do the same.

"I would hope he would put his share of the settlement into the trust fund also," Tomblin told me. "He said he would check with his people and let us know, but he hasn't let us know yet."

The attorney general's office would get an $8 million cut. A Morrisey spokesman wouldn't say what it would do with the money.

The opioid epidemic had touched Tomblin personally. His younger brother was an addict. In 2014, Carl Tomblin was arrested after selling prescription painkillers to an undercover informant. Governor Tomblin released a statement saying he was "saddened by my brother's actions, and I am disappointed in him, but I love him." Carl Tomblin pleaded guilty. A federal judge sentenced him to eight months of home confinement. He started taking Suboxone, a prescribed medication for opioid dependence, putting him on a path to recovery.

Cardinal agreed to pay the state $20 million, and AmerisourceBergen would toss in $16 million. The companies denied any wrongdoing. Previous settlements with nine smaller distributors netted an additional $11 million. I asked Tomblin whether $47 million was enough to pay for the misery inflicted on the state. "The amount of money, and the grief and loss of life that these pills have cost the people of West Virginia, it's hard to put a dollar figure on it," he said. "The treatment is going to have to go on for years."

I also telephoned Cagle. He said the state got the best deal possible given the circumstances. A mock trial with a focus group weeks earlier hadn't gone as well as expected. Potters might have sunk the state's case. The new Supreme Court majority promised to be more "pro-business." There were too many wild cards. The $36 million settlement made sense. The outside lawyers—Cagle and Jackson among them—would collect about a third of the settlement proceeds.

Under the terms, Cardinal and AmerisourceBergen agreed to promptly alert state authorities when they spotted suspicious drug orders. The settlement specified that the reports be sent to state police headquarters and the attorney general. Morrisey's and Tomblin's offices issued a joint news release later that day. The drug distributors had a say in the wording of the release—that also was part of the settlement. The statement emphasized that the companies shipped drugs only to licensed pharmacies. The distributors used the same language in their own releases.

The settlement closed out the four-and-a-half-year fight. The distributors were pleased to put the allegations behind them. They could direct their full attention back on delivering drugs to patients who needed them. But their legal troubles weren't over—far from it.

Kanawha County Commission president Kent Carper, representing West Virginia's most populous county, quickly mailed a letter to the defense firms representing a dozen drug distributors, including the Big Three—McKesson, Cardinal, and AmerisourceBergen. Carper put the companies on notice that the settlement with West Virginia's attorney general wouldn't stop towns and cities and counties from filing their own lawsuits. They had the right to fight back, too.

PART IV

22

List of the Dead

The trail of painkillers led me to Kermit, to Debbie Preece's living room, and a sheet of crumpled notebook paper she pressed into my hand. It was my first visit to Debbie's house, the first time we had met. Names were on both sides of the lined sheet: Fig, Okey, Homer and Drema, Roxy, Little and James, Tim, Lynn and Gary. She'd scribbled brackets around some names to show family relations. A husband and wife. A brother and sister. Three Maynards had died.

It was late January 2017, and I was on the edge of Debbie's sofa, counting the names with her brother Tommy "Tomahawk" Preece. His parents gave him his nickname after he was born in the back seat of one of their taxis in Tomahawk, Kentucky, on the way to the hospital. My story about huge shipments of opioid pills to West Virginia and its rural communities had hit like a thunderclap, triggering towns and cities across the state to file lawsuits against the drug distributors. Kermit was one of the first to do so. Earlier in the day, I had followed the mayor and town council members to the Mingo County Courthouse, where they hand-delivered the lawsuit to the circuit clerk. They sought to recoup the town's costs in dealing with opioid abuse.

Kermit mayor Charles Sparks showed me the town's ledger back when Sav-Rite was raking in millions of dollars each year. It was the first time I had seen it. Jim Wooley reported minimal sales and paid only a few thousand dollars a month in business taxes. There was no telling how much he had stiffed the town's coffers. "To dump that many pills in a little ol' town like this," Sparks said, shaking his head. "It's mind-boggling what was going on."

I had arranged to meet Tomahawk at the fire station and then followed him across the railroad tracks to Debbie's house. Jim Cagle had suggested I talk to her. The television was tuned to a late-morning talk show. Debbie pulled her feet up into a recliner. She explained the list of names.

"At least a year ago, me and my friend were just talking, and we said, 'I wonder how many there is around here.' I said, 'Let's just make a list of who we know.' We knew for a fact that they had OD'd, so we just started writing names down."

They remembered fourteen from Kermit, another dozen or so from across the waters of the Tug Fork in Martin County, still more from nearby Crum. All had lived within a twenty-mile radius. They had traded hugs and handshakes at the IGA, Dairy Queen, and Heilig-Meyers furniture store in town. They picnicked in the park at Big Bottom, before discarded syringes started showing up on the playground. With her friend, Debbie jotted down name after name, stopping only when memory faded. She kept the list in a dresser drawer upstairs, beside photographs of Bull—and his empty prescription bottles. This was the first time she had set eyes on the list of names in months. "None of these are heroin," she said.

"All of them are prescription pills," Tomahawk said. He had been chasing down fires since he was five years old, his father hoisting him up into the cab of the fire engine, letting him ride shotgun, siren wailing. Now, there were few emergency calls for blazes—most were for overdoses.

"If this is just the people we know, the people from this immediate area," Debbie said, "well, can you imagine what's going on all over the county, let alone the state and everywhere else?"

Mingo County had the fourth-highest prescription-opioid overdose death rate in America, while neighboring Martin County ranked fourteenth. Debbie and her brother were surprised both counties weren't even higher.

"There's quite a few," Tomahawk said. "If you could find all of them, I think you would be shocked. You see it, and you wonder what's happening to our people, what's happening to our community. People are OD'ing left and right."

We were counting the names again, starting over from the beginning. The list had no order—just people's names and where they had lived. Bull Preece's name was on the list, near the middle of a second column. No marks distinguished his name from all the others. More people had died in recent months, but Debbie hadn't yet added them.

"That's sixty-one right there," Tomahawk pronounced, handing the sheet to Debbie. Sixty-one overdose deaths. All close by. "I'd say if she'd do another one of these, you'd need more than one sheet of paper."

"We'd have to have a book, Hawkie," Debbie said.

She wanted to fix a problem but didn't know how. She had no children, but she worried about her nieces and nephews—there were probably fifty. What was their future in a town overrun by drugs? Good-paying jobs running coal were increasingly hard to come by. She had started something—a crusade, that's what it was—twelve years earlier, but what had she accomplished? The pharmacies were back in business with many of the same employees, the doctors who went to jail were now operating addiction treatment centers, the distributors were still reaping enormous profits, and people were overdosing in record numbers.

"Everybody around here is doing pain pills," Debbie said. "There's nowhere to hide, nowhere to run. Everybody's looking for some way out of this mess. You don't want your children, your loved ones, to leave here. You certainly don't want them to go out of here in a pine box."

For nearly two hours, she'd answered my questions—now, she had some to ask. "Eric, do you have the answers? Do you have solutions? Do you have ideas yourself?"

I stammered, telling her state lawmakers had earmarked money for treatment beds. The opioid epidemic was starting to draw attention from the national media. I didn't know what else to say.

"I'm starting to hear some solutions," I said, "but—"

"Bull taco," Debbie said.

The house shuddered as a Norfolk Southern train blared by, steel on steel, horn shrieking. The noise startled me. Debbie didn't flinch.

"With the OxyContin, there's something about that pill," she said. "There's just something about it that's so addictive. There's something in these pills."

I asked about Sav-Rite. I knew of the pharmacy only from its mention in the state lawsuit and the startling pill numbers on the DEA spreadsheets. I had driven through Kermit on an assignment years ago. This was the first time I had reason to stop.

"There were all kinds of people," Tomahawk said. "Just too many license plates coming from everywhere—North Carolina, South Carolina, Virginia, Ohio."

"They'd get free popcorn," Debbie said.

"Like at the fair?" I asked.

"There was like a little camper trailer," she said. "It was hot dogs and hamburgers and stuff."

"Like with a little hitch on it," Tomahawk said. "They weren't able to keep it there long, I don't think."

"They started getting heat," Debbie said. "There were people coming in filming and watching. I think they got the message."

"Who's got the film?" I asked.

"Mr. Cagle," Debbie said. "James Cagle."

I'd have to pay a visit to his office and see if he'd let me borrow a copy of the surveillance video. Debbie warned that Cagle wasn't savvy with technology. I might have to ask an office assistant for help finding a way to play the tape. I imagined myself trying to spool a film projector.

"Have you known Mr. Cagle long?" I asked, making small talk.

"Thirtysome years," Debbie said.

"How did you know him at first?"

Debbie glanced at Tomahawk. He shook his head, but said nothing. I sensed something was wrong. I hadn't intended the question to stir suspicion.

"Off-record?" she asked.

"OK. Off-the-record." I switched off my digital recorder and closed my pocket-size spiral notebook.

Debbie told me the story of Kermit past, of drug deals from a double-wide and bags of pot thrown from trains. Tales of federal raids and the family business and time behind bars. Debbie's past, which she'd just as soon leave there. She'd hidden this dark, painful chapter from the youngest generation of Preeces.

23

Damage Control

The Drug Enforcement Administration had a public relations problem. How could millions of pain pills end up in small towns in West Virginia if the DEA was doing its job? In mid-February 2017, just after the lawsuit settlement, it dispatched top administrators from its Washington field office to Charleston. They planned to announce a $500,000 drug-fighting initiative called DEA 360. A spokeswoman inquired whether I'd be interested in interviewing Special Agent in Charge Karl Colder the afternoon before the announcement. I was interested.

Colder greeted me at the Marriott hotel. Four other agents were sitting at the conference room table. My digital recorder was rolling. I asked if they knew Kyle Wright, the DEA agent whose emails to Morrisey's office had quantified the opioid pill shipments to West Virginia. "Oh, I know him," said Special Agent Ruth Carter. "He works for me."

Reporters and lawmakers from Alaska to Maine had been calling me at the capitol pressroom for weeks, asking how they could get the same DEA data, the same pill shipment numbers for their states. I referred them to the DEA. I wanted the information, too. It was a reasonable request.

"You're wasting your time," Carter said. "All you would get

191

would be a bunch of redacted documents. I mean, it's to protect the company."

What about protecting the public?

"It's proprietary information," she said. Judge Will Thompson and the West Virginia attorney general's office had made a different determination.

In late 2016, *Washington Post* reporters Lenny Bernstein and Scott Higham published a series of explosive stories about the DEA's cozy relationship with drug distributors. The DEA, in the face of pressure from the drug-shipping industry, had dramatically slowed an enforcement campaign designed to hold distributors responsible for excessive sales of painkillers. The "closed system" that the distributors and their lawyers liked to tout was leaking, and the DEA was turning a blind eye, the *Post* revealed. I expected the agents in the room to dispute the coverage, but instead they confirmed that the Department of Justice was, as reported, tying their hands.

"So we were also targeting their DEA numbers, their registrations," Special Agent Carter said, "but unfortunately, if you read the *Washington Post*, you'll know, unfortunately, there was some pressure and we were told—"

"That's true?" I interrupted. I needed to learn to keep my mouth shut while people were talking. It was a bad habit.

"If Joe Rannazzisi said they said that to him, I believe they said that to him," she told me. Rannazzisi, the now-retired DEA official who testified during the 2012 congressional hearing on prescription drug abuse, had spoken to the *Post* for the articles. Carter stressed that she wasn't in the room when Rannazzisi was ordered to stand down, but she had felt the same pressure. "One DOJ official told us we could not push Cardinal any further," she said. "That's the only thing I know that's true."

"So it was the DOJ's people, not the DEA's?"

"At the DEA, we want to do the right thing," she said.

While an unprecedented number of Americans were dying of drug overdoses, as Carter told it, DEA agents were told to call off their investigations into some of the nation's biggest pill distributors. Some of the DOJ lawyers, and DEA lawyers as well, had represented the distributors while in private practice. Meanwhile, the distributors would hire their own attorneys and drug compliance officers, some of whom had worked at the DEA. They were making $500 an hour to stroll through this revolving door—and few seemed to give a damn that pain pills were killing people.

"And when taking an action against the distributors, it's just going to be more complicated," Carter said. "It's going to be harder because you're dealing with very high-priced attorneys. They'll hire three or four law firms to represent them."

"The revolving door," I said.

"The carousel," Colder said. "And I have no problem with that. They're going to pay the big dollars, and whoever has the experience, that's who they go after. You have to look at it positively. Hopefully, they're going to train people how to do things the right way."

Colder's candor surprised me. He had taken over the DC field office in 2013. The DEA pill numbers we obtained had covered 2007 to 2012. I asked why the DEA hadn't been watching what was going on in West Virginia.

Carter answered first. "We had two diversion investigators in the whole state of West Virginia. So they had access to that data, but if there's only two of them to cover the entire state, it's a little hard when you've got certain things you have to do." What things? I wondered. The agents' duties included scheduled inspections of drug treatment programs, she said. Since my article about West Virginia's being showered with painkillers, however, the DEA had hired three more investigators and appointed an "assistant special agent in charge" who reported directly to Colder. More agents were on the way.

"You've had no leadership in West Virginia, OK?" Colder said. "And this is the first time, maybe, the community has seen the special agent in charge. This is the first time."

The next day, I covered the press conference. We published a story about the DEA's recent efforts to crack down on street drugs such as heroin and fentanyl. For Sunday's paper, however, I wrote a second article. The headline read, "'We Had No Leadership' in WV amid Flood of Pain Pills." A DEA spokeswoman called the next day. She was furious. Didn't I realize the entire conversation at the Marriott was off-the-record? What was I doing quoting special agents?

"Don't even ask," I wanted to say. Instead, I reminded her of the recorder I had placed in the middle of the table in plain sight, for everyone to see. I offered to play the tape for her. Fifty-two minutes into the interview, Carter asked to go off-the-record to talk about why the distributors started sending reports about questionable drug orders to the Board of Pharmacy. I honored Carter's request. I didn't use those quotes in the article. I considered the rest of their comments fair game. Besides, if they wanted me to turn off my recorder, all they had to do was ask.

Towns, cities, and counties were filing lawsuits right and left against the drug distributors, so AmerisourceBergen sent lawyers and executives to Charleston in early March 2017 to suggest ways to make my reporting, in their words, "more informative and engaging for readers." Instead, I received a lecture on what I had neglected to cover. It wasn't the spate of lawsuits.

Gabe Weissman, the company's vice president of communications, invited me to bring my "team" to the sixteenth-floor conference room at Jackson Kelly's law offices. I had no team, but Rob Byers, the newspaper's executive editor, offered backup. Al Emch would do most of the talking for AmerisourceBergen, even though the executives and in-house lawyers had jetted in from Philadelphia, where the company was headquartered. I mentioned I grew up in the suburbs of Philadelphia. Emch took note of that and started a history lesson about the American Revolution. I assumed he was trying to break the tension.

I wasn't sure what point Emch was trying to make. Perhaps he was hoping to lighten the mood. Or maybe it was a lesson about overcoming the odds. I thought of a different parallel, an invading force that unleashed a poison on a vulnerable population. In the sixteenth century, Swiss chemist Paracelsus coined the adage "the dose makes the poison." It's a fundamental principle of toxicology: High concentrations of any substance can kill. In this case, prescription opioids were particularly deadly. The coal barons no longer ruled Appalachia. Now it was the painkiller profiteers.

"I do want to mention one other thing along the lines of the rules of engagement," Emch said, apparently having slipped during his digression into a martial frame of mind. He then rattled off all the stories I hadn't written. I hadn't reported that the DEA met with drug manufacturers every year and set quotas on how many controlled substances could be sold in the US market. I hadn't reported that legitimate patients—those with cancer and in hospice care, for instance—weren't able to find doctors who would prescribe them pain medication. What's more, I had neglected to write about the DEA not allowing one distributor to see what another was shipping to the same pharmacy.

"We haven't read much yet about that," Emch kept repeating.

It was a good thing I'd brought along my team to switch the subject. "You mentioned about all the data that you're not privy to, the things you can't see," Byers said, "but what about what you can see, your own numbers skyrocketing?"

"*Skyrocketing* is a pretty strong term," said Emch.

I had some of those numbers with me—borrowed from Kyle Wright's spreadsheets that he'd sent to the attorney general. AmerisourceBergen's oxycodone sales had nearly doubled in West Virginia over six years. The distributor shipped eighteen times as many oxycodone pills to Mingo County in 2012 as it did in 2009. Other counties also had huge increases.

Emch had to be careful. He didn't want to give away anything that would help the trial lawyers suing his clients. And he was trying to stop other cities from doing the same. Over and over, he lauded AmerisourceBergen's failproof program for identifying and blocking suspect orders from individual pharmacies.

"I will say to you, in our view, it is the best in the industry," he said.

Byers persisted. "I'm referring to year-over-year data, rather than twenty-four-hour data or suspicious orders. You have this fantastic system here. It must show this increase."

"When an order goes outside our parameters, then it's kicked up for further review."

"So you really don't look at your yearly numbers?"

"Don't look at what?" Emch asked.

I asked the same question. Did the company analyze yearly county and statewide sales of painkillers? That's all we wanted to know. The interview was diverting from AmerisourceBergen's talking points.

Weissman stepped in. "I don't want to be coy and say I don't understand what you're asking us . . . but when you look at year-over-year results, this category of products we distribute is driven by prescriber behavior."

They were back on message. It was the doctors' fault. "There was overprescribing," Emch said. "There was out-of-control—"

"If you knew that about the doctors," Byers said, "then shipping these high numbers of pills—"

"Rob, you're on the wrong track," Emch said.

Rob wasn't on the wrong track. It just wasn't the track the distributors wanted us on. Weissman, a seasoned and effective corporate spokesman, started talking again about DEA quotas, which set the number of pills that could be manufactured each year. We were back on track.

I swiveled in my chair. I had a specific example to support our question. It was from Berkeley County in the state's Eastern Panhandle. Berkeley had the second-highest drug overdose rate in

West Virginia the previous year. "What we're trying to say, when a county goes from 168,000 pills, and five years later, you guys are selling 1.1 million—" I started.

"We're dealing with individual customers," Emch interrupted. "If there was something that popped up to look at, we looked at it—increases or because legitimate demand was increasing."

"I think *legitimate* is the key word," Byers said. "Wouldn't that be a red flag for you to see that kind of increase?"

The AmerisourceBergen executives were keeping a watch on the time. They had a return flight to catch to Philadelphia in less than an hour. They had wheeled their suitcases to the conference room.

"The red flags that exist out there are red flags that are embedded in our system," Emch said. "And they're not red flags about the nationwide or statewide oxycodone prescriptions that have doubled in the last two years. It is, again, red flags that are focused on each individual customer that we have and every order that they submit. That's the way the program works."

We hadn't asked about oxycodone prescriptions, just pills. But, if I understood Emch correctly, this best-in-class program for keeping pain pills off the streets wasn't tracking monthly or yearly sales by county or state. It wasn't stopping any unexplained aggregate increases.

The meeting broke up after two hours. Rob and I weaved our way through downtown back to the newsroom. Rain was misting down. Rob, an Eagle Scout, snapped open an umbrella. I had a lot to sort out, but the same question kept swirling in my mind. I should have asked the question, but didn't think to at the time. The computer software program didn't flag the annual spikes, but surely the company's sales managers knew what was going on. They knew pharmacies' orders for pain pills were escalating. They saw the sales numbers. So why didn't they say anything? Why didn't they sound an alarm?

• • •

By the summer of 2017, Cardinal Health was playing hardball. It was determined to stop the onslaught of lawsuits by any means necessary—including by trying to discredit my reporting. The stakes were higher now. It was one thing to have a small town such as Kermit suing your business. It was another to have the state's largest cities—Charleston and Huntington and Parkersburg—doing it. Cardinal was sending its lawyers to city council and county commission meetings across West Virginia to pressure public officials to drop lawsuits or to stop them from filing them in the first place. However, cities and counties in other states had also started suing the distributors. Things were getting out of hand.

In August, an email arrived in my *Gazette-Mail* inbox from the lawyer representing the city of Huntington. No city in America had been hit harder by the opioid epidemic. Huntington was the new ground zero. The city had made national headlines the previous summer after twenty-five residents overdosed within five hours. The overdoses were linked to a bad batch of heroin laced with fentanyl, a synthetic opioid fast becoming the leading cause of drug overdoses.

Attached to the email was Cardinal's latest response to Huntington's lawsuit, a "notice of nonparty fault." This legal maneuver was designed to spread responsibility to others and was one of the most effective weapons in a defense lawyer's arsenal. Cardinal identified more than two thousand entities that could be "wholly or partially at fault" for diverting prescription opioids for illegal use. The list named doctors, nurses, convicted drug dealers, law enforcement agencies, hospitals, ambulance services, the mayor's office, fire departments, health insurance carriers, pharmacies—and Lily's Place, a Huntington clinic that nursed drug-exposed newborns through withdrawal. I grabbed the phone in the capitol pressroom and dialed Huntington's lawyer, Rusty Webb. Cardinal's filing was pointing a finger at a nursery that cared for babies with neonatal abstinence syndrome. A nursery, even one with a license for administering controlled substances, couldn't be held responsible for fueling the drug crisis. It was inconceivable.

"It's as dirty as hell," Webb said.

Cardinal's spokeswoman had an entirely different take. She described it as a procedural step. Huntington's lawsuit essentially implicated all doctors, pharmacies, and clinics licensed to dispense pain medications in the area. Lily's Place had a license to dispense methadone and other controlled substances used to wean babies off opioids. It was up to Webb to tell Cardinal who diverted opioids for illegal use and who didn't. Until then, the list would remain unchanged.

"The plaintiffs have not identified anyone, but rather just put up a total number of pills," Cardinal's spokeswoman told me. "By doing so, they implicate everyone who handled every pill regardless if it went to a cancer patient or some other legitimate medical need."

Still, I wondered if Cardinal's lawyers bothered to review the list of licensees. Wouldn't they have realized naming Lily's Place would reflect poorly on the company? Lily's Place had drawn national attention for its lifesaving work with newborns and their mothers. The freestanding clinic handled the overflow of babies from area hospitals' neonatal units. Politicians wanting a photo op frequently stopped there. I called Rebecca Crowder, who managed Lily's Place. She was as surprised as anyone else that the neonatal clinic would show up in Cardinal's filing. She told me she wanted to weigh her words carefully. She didn't want to pick a fight with a giant drug company. She didn't want Lily's Place used as a pawn by either side.

"This proves that no research has been done and that names are being thrown out as a technique to deflect blame," Crowder said.

The next day, we published a story about Cardinal citing Lily's Place for possible fault in the opioid crisis. Politicians were outraged. Congressman Evan Jenkins, who helped start Lily's Place, called Cardinal's legal tactics "shameful." Senator Joe Manchin said he was "shocked and appalled" Cardinal would have the gall to blame Lily's Place. My article revealed that Cardinal's filing was made possible by a new law that Republican legislators hailed as "tort reform," which would make West Virginia friendlier to businesses. The law was copied from legislation recommended by the American Legislative Exchange Council, an industry-backed group known for pushing conservative policies nationwide. State lawmakers had given the drug distributors a new tool to fight the lawsuits being filed by cities and counties.

It didn't take long for Cardinal to fight back. The spokeswoman submitted a letter to the editor two days later. We published it in the Sunday paper. It began, "Eric Eyre's Aug. 9 article presents egregiously false and irresponsibly cherry-picked claims about Cardinal Health's proactive approach to the opioid epidemic." The rest of the letter reiterated Cardinal's statements that I had incorporated into the story. The company was repeating its rationale, promising to consider removing Lily's Place from its list if Webb would admit that the neonatal clinic didn't order prescription drugs that found their way into the hands of drug abusers. The spokeswoman wrote, "Cardinal Health cares deeply about the devastation caused by opioid abuse and is leading the way to address the epidemic."

It was Cardinal's first public criticism of one of my stories, but it wouldn't be the last time the company would try to dilute its share of responsibility for the opioid crisis. In a separate court filing, Cardinal identified 334 health professionals and organizations in McDowell County, West Virginia's poorest county, as potentially at fault. The list included a dead man, Dr. Ebb "Doc" Whitley, who perished in 2005 after a neighbor set the doctor's house on fire. Cardinal also named Welch Community Hospital, the only hospital in the county. The distributor had managed the hospital's pharmacy for twenty years. The pharmacy's employees worked for Cardinal.

• • •

As Cardinal was digging in against its multiplying litigants, it and the other big distributors were about to come under heavy fire on a whole new front. I found out by way of an email from Jennifer Sherman, press secretary for the US House of Representatives' Committee on Energy and Commerce. Its staff members had been following my reporting on the huge number of opioids in West Virginia, and now the committee was offering me an exclusive. "I think you might be interested in some action we'll be taking," Sherman wrote.

It wasn't every day that Congress contacted our newspaper with a tip. It would be akin to West Virginia lawmakers reaching out to the *Washington Post*. I was familiar with the House committee's chairman, Representative Greg Walden, R-Oregon. In 2010, I reported that West Virginia officials had used $24 million in federal stimulus funds to buy giant-size Internet routers designed to serve entire colleges and instead installed them at small-town libraries. Walden demanded investigations and hearings. Auditors concluded the state had wasted millions of dollars. Now, Walden's committee had taken an interest in my opioid coverage.

Within an hour, I had Sherman on the phone. The House committee had just started an investigation into the distributors' pain pill shipments to West Virginia. It was sending letters to the CEOs of Cardinal, McKesson, and AmerisourceBergen, directing them to turn over records that would show the number of hydrocodone and oxycodone pills sold to pharmacies each year from 2005 to 2016. Federal lawmakers wanted the names and addresses of warehouses that served West Virginia. The letters also asked the distributors to answer questions about their drug sales tracking systems and policies and procedures, and whether they referred questionable orders for prescription painkillers to the DEA and West Virginia Board of Pharmacy. The committee cited my articles and asked the distributors whether my numbers were accurate.

"If these reports are true, it would appear the state of West Virginia may have received extraordinary amounts of opioids from distributors beyond what that population could safely use," Walden wrote. "The possible oversupply described in this reporting suggests that such practices may have exacerbated the opioid addiction problem facing the state." Four other committee members, including David McKinley, R-WV, also signed the letters.

I phoned the three companies for comment. Weissman said the DEA already had the sales data sought by the House committee. The DEA got a letter, too. The panel asked for the same hydrocodone and oxycodone numbers, and for an explanation of what the agency had done to curtail the flow of highly addictive painkillers into the state. Sherman told me the committee was also investigating the DEA's decline in enforcement actions, citing reports in the *Washington Post*. The DEA and the distributors were given thirty days to hand over everything.

24

A Death in Marrowbone Creek

It was the worst 911 emergency call he had ever answered. On a Wednesday afternoon in August 2017, Tomahawk Preece's cell phone squawked—the whoop-whoop of a siren, followed by static and the dispatcher reporting a possible overdose at Marrowbone Creek. Tomahawk was at home, and in one swift motion he tamped out a cigar, grabbed his keys, and burst through the door. He briefly considered stopping at the firehouse to pick up a helping hand, but he thought the better of it. With an overdose, time was everything. And the dispatcher's voice had been urgent. He gunned his rescue truck up the two-lane toward the creek. It took him ten minutes to get there.

At the bottom of the hollow, he saw a woman standing on the gravel road. No, she was jumping up and down, waving her arms, screaming. He recognized her instantly: Timmy Dale's girlfriend. He knew right then it was his brother who was dying and he couldn't get to him fast enough.

"Boys, it's one of ours," Tomahawk shouted into his two-way radio. "Get here quickly."

Timmy Dale Preece, assistant chief of the Kermit Volunteer Fire Department, was slumped over the handlebars of his four-wheeler. Tomahawk grabbed his barrel-chested brother in a bear hug, pulling him to the ground. He wasn't breathing. Tomahawk administered CPR and then reached for the defibrillator. He ripped open Timmy Dale's shirt, stamping the electrodes on his chest. The machine detected a heart rhythm, which meant Timmy Dale might be alive, then delivered a shock. There was no response. Tomahawk waited for the second hit. It never came. The defibrillator detected nothing. It wouldn't shock. Timmy Dale had flatlined. He was dead at fifty-two.

As a volunteer firefighter—more than four decades if you counted the time he'd started racing to fires with his father—Timmy Dale had saved dozens of lives in southern West Virginia and eastern Kentucky. Responding to car wrecks, he could work the Jaws of Life better than anyone else in the county. He'd cut you out of a vehicle in no time, the pincers snapping through the steel hulk like scissors to tinfoil. Back at the firehouse, he would amaze the other volunteers. He'd set a milk crate on the floor, place a cup upside down atop it, and balance an egg on the cup. Like a magician, he would heave up the weighty Jaws of Life and pluck the egg off one cup and set it atop another, back and forth, never squeezing the shell so tight that it would crack, and never so loose that it dropped to the concrete floor. The young volunteers would watch Timmy Dale in awe.

When he wasn't rescuing people from mangled vehicles, he was diving into the Tug Fork of the Big Sandy River, pulling them out of high waters. Timmy Dale was lead diver for Kermit's swift-water search-and-rescue team. He'd motor out into the raging waters in an aluminum johnboat, strap on a mask and flippers, and be the first to plunge in. The summer before, Timmy Dale got a call to help flood victims in Greenbrier County, a three-hour drive from Kermit. Floodwaters were hurling people's homes around like flotsam, and residents in the town of Rainelle were trapped. Timmy Dale never hesitated. At 1:00 a.m. the next day, with the water still rising, he and his four-member team started plucking people from rooftops. The Kermit crew rescued twenty-five over the next couple of hours. Swift-water rescue squads with more men, newer boats, and more expensive gear marveled at the speed and courage of the boys from Kermit. Timmy Dale had trained them well.

But now, Tomahawk and Kermit's other volunteer firefighters were gathered around the body of their assistant chief at Marrowbone Creek, and there was nothing they could say. Their blond-haired, blue-eyed hero, the man who had taught them everything about saving lives, was gone, and they couldn't save his. The cause of death wasn't immediately known. Because he overdosed, there would be an autopsy. The sheriff's deputies questioned Timmy Dale's girlfriend. She admitted they had been snorting cocaine and meth, and dealers in the area had started lacing those drugs with fentanyl. Timmy Dale had started vomiting. He was dizzy, staggering. He collapsed as he tried to turn the key in the ATV's ignition. They would later find five roxies (generic oxycodone tablets) in Timmy Dale's pants pocket. Tomahawk told a local TV station his younger brother had died of an apparent heart attack.

Debbie was devastated by Timmy Dale's death, a second younger brother lost to a drug overdose. She and Tomahawk had a bad feeling that more could have been done. The fire department in Williamson had an overdose-reversing drug that would snap people awake after they'd stopped breathing. It was naloxone, but everybody called it by its brand name, Narcan. Kermit and the other volunteer fire departments in Mingo County didn't have it. They couldn't afford it. Firefighters were responding to overdoses without a critical lifesaving medication. It might have saved Timmy Dale. Access to a drug that saved lives wasn't available, while access to the drugs that killed was plentiful. That wasn't right, said Debbie, who called me, asking if I could do anything to help the town get Narcan. I promised her I'd do a story about it, shine a spotlight on the problem.

The funeral parade for Timmy Dale was fitting for a fallen fireman. The Belfry and Williamson fire departments had brought their ladder trucks and hung a giant American flag over Highway 52, the main drag through town. Kermit's firefighters, wearing their navy dress uniforms, white gloves, and caps marched in two lines along the hot pavement. One carried Timmy Dale's helmet. A wooden casket, enveloped by a smaller American flag, rested atop the department's white fire engine. Tomahawk squatted beside it on the truck's roof like a watcher of the sky in the bow of a ship. Firefighters and paramedics from other companies saluted Kermit's brigade. The procession turned west, crossing the bridge over the Tug Fork to the cemetery. A line of slow-moving fire trucks followed. Their motors rumbled, emergency lights flashed, brakes squealed. There must have been fifty at least. Behind them, a convoy of cars and trucks stretched for more than a mile. Hundreds of mourners had come to pay their respects.

A 911 dispatcher's voice cut into the emergency radio frequency, and those in the procession idled their vehicles to listen over the airwaves. The female voice was paging Timmy Dale, a last dispatch call, the end-of-watch call. She twice called out his name and engine number. There was no answer. Static crackled over the next few words.

Having heard no response from Assistant Chief Timmy Dale Preece . . . he has responded to his last call. Assistant Chief Timmy Dale Preece served the Kermit Fire Department and the citizens of Mingo and Wayne Counties for over forty years. We appreciate and honor Timmy Dale's dedication and sacrifice during this time. Assistant Chief Preece has now become a guardian. He will look over all firefighters as they respond. Assistant Chief Timmy Dale Preece has completed his tour here on earth. Rest in peace, brother; we'll take it from here.

By late October 2017, Oregon congressman Greg Walden had lost his patience with the high-ranking administrators at the Drug Enforcement Administration. He was fed up with their dodges and delays. What was it—five, six months—since his committee had asked the DEA to turn over records that might help everyone understand the opioid crisis? More people like Timmy Dale Preece were dying every day. More families were being ruined. Why wasn't the DEA cooperating? Now was Walden's chance to demand an explanation.

Neil Doherty, assistant administrator of the DEA's Office of Diversion Control, was on Capitol Hill, testifying under oath before Congressman Walden's committee. Walden directed the proceedings from the dais. After Doherty gave his opening statement, Walden started with what he called a "simple question": "Which companies supplied the pharmacy in Kermit, West Virginia, that received nine million opioid pills in two years?"

"Thank you for that question, Chairman," Doherty said, shuffling papers at the witness table. "Currently, we are reviewing the request from the committee, and I do not have that data with me today. I apologize."

Walden wasn't in the mood for apologies. He glanced down at Doherty like a school principal listening to a student trying to explain bad behavior. "So we have asked for this information in a meeting. We have asked for this information in an email. We have asked for this information in a letter, and we have asked for this information now in a hearing." Someone's cell phone jingled, and Walden asked everyone to switch their phones to silent mode. "We have been trying to get to the pill-dumping issue in West Virginia for a very long time." Walden rubbed his forehead. "To me, this is a pretty basic question: Who are the suppliers?"

Walden threatened to send a subpoena to the DEA. Doherty apologized again. There would be no more excuses. The DEA would try to expedite answers to the committee. But Walden wasn't finished. "Enough is enough." He leaned forward with a stack of documents clutched in his hand. "Just for members' awareness on both sides of the aisle, the committee received yesterday a set of documents from an anonymous source. Bipartisan committee staff are now reviewing these documents." Walden didn't elaborate, but his next question hinted that something nefarious was going on at DEA headquarters. "Mr. Doherty, have you or anyone at the DEA that you are aware of received any instructions or directives to erase emails or otherwise destroy documents on this matter or any others?"

"No, sir. I'm not aware of that, nor have I been in any conversation relative to that matter."

The DEA's delays hadn't stopped the staff members on Walden's committee from doing their own research on pill dumping. They examined drug-shipment data, broken down by the three-digit ZIP code prefixes in West Virginia. Hydrocodone shipped to the Mingo and Logan counties ZIP code area increased 600 percent over eight years. Meanwhile, oxycodone shipments to the Wyoming and McDowell counties prefixes doubled over five years. Diana DeGette, a congresswoman from Colorado, quizzed Doherty about those findings. "Would you agree that some of these trends are troubling?"

"Yes, ma'am, I would."

"Do you know who did this?"

"It is my understanding currently that we have information relative to companies involved, and we are reviewing that data to determine what we can legally—"

"And I assume we will get that answer?"

"Yes, ma'am."

After breaking for lunch, Congresswoman Kathy Castor of Florida had her five minutes to question Doherty. Most of her colleagues hadn't returned. She asked why the DEA had failed to flag the massive orders of prescription opioids in southern West Virginia.

"Ma'am, thank you for that question. And DEA agrees the amount of pills going into that area was excessive in looking back." Doherty dropped his guard a bit. He admitted the DEA wasn't paying much attention to West Virginia in the early days of the crisis. Pain clinics and pill mill pharmacies were popping up in the Democratic congresswoman's home state. The DEA was all hands on deck in Florida. "Florida was the epicenter of the beginning, in some ways, of the opioid crisis that we face today."

So nobody at the DEA was monitoring painkiller shipments into particular communities elsewhere? Castor asked. Couldn't someone at headquarters have done that?

"Ma'am, the way these are monitored in conjunction with distributors, they are monitored through the submission of suspicious orders. And the distributors have an obligation to report that to the DEA, and that was a flaw and that is why." The DEA was relying on distributors to report suspect pharmacies, but the companies weren't going to rat on their customers. The distributors had a disincentive to turn anybody in. They were in business to sell drugs, not to police pharmacies.

"So clearly, there is a breakdown here," Castor said.

25

"How in God's Name?"

The chairman of Miami-Luken wanted to talk to congressional investigators in private. Dr. Joseph Mastandrea had agreed to the sit-down interview. The other distributors' executives had declined the same offer. On a Wednesday morning in mid-December 2017, Mastandrea and his lawyer walked into the Ford House Office Building in Washington. Alan Slobodin, chief investigative counsel for the House Energy and Commerce Committee, asked Mastandrea about his background.

"My position is due to nepotism," he said.

Mastandrea's father had started the company, and now his son was overseeing its collapse. The DEA was trying to revoke Miami-Luken's license. A drug distributor that didn't deliver controlled substances didn't stay in business for long. It would lose customers in droves. Painkillers were profitable. Mastandrea was fighting for the company's survival. He first got wind of trouble in 2013. The DEA was sending subpoenas to Miami-Luken's headquarters in Ohio. Mastandrea hired a Washington-based lawyer. They summoned the CEO and chief compliance officer. The lawyer reviewed Miami-Luken's sales history. Mastandrea told investigators that he was as surprised as anyone.

"He brought to my attention the inordinate amount of product we were selling," Mastandrea recalled. "I was apoplectic. I informed everyone in the room I would rather shut the expletive down than allow product to be diverted."

But Mastandrea didn't shut down anything. Miami-Luken kept shipping opioids. He blamed the CEO, a former railroad executive. He blamed the chief compliance officer, a former credit manager. He blamed practically everybody but himself. He explained he was a full-time doctor. He didn't stick his nose in day-to-day operations. He trusted others. But he now acknowledged that the company had done next to nothing to identify questionable, high-volume orders from pharmacies, though it had started a drug monitoring system as early as 1995. "Unfortunately, these efforts were primarily based on one's feeling about what constituted a suspicious order," he said.

Slobodin, a twenty-year committee lawyer, did a double take. "What do you mean by that?"

"It was one individual's feeling about whether or not this order represented an unusual quantity, frequency, or whatever the Controlled Substances Act says."

Miami-Luken hadn't set any daily or monthly limits on the number of painkillers that pharmacies could buy. It was going on someone's intuition about whether too much was too much—the intuition of a CEO who received bonuses each year while profits soared. The company tried to get a handle on things, Mastandrea said. At his direction, it bought a sophisticated software program to flag suspicious orders. But it sat idle for two years. Nobody knew, or bothered to learn, how to run it. At one point the program was identifying every single order as suspicious. The compliance officer eventually shut it down.

"I was shown the system," Mastandrea said. "Unfortunately, I didn't know what I was looking at. And even more unfortunately, neither did our compliance officer."

The congressional committee had directed Miami-Luken to turn over its files on Sav-Rite Pharmacy. At first, the company provided a single page that tallied the 5.63 million hydrocodone and 94,600 oxycodone pills shipped to the drugstore between 2005 and 2011. A committee staff member handed Mastandrea that document and a printout of the US Census Bureau's population estimate for Kermit.

"So these two documents together—do you believe that this was an inordinate number of pills?" Slobodin asked.

"It certainly looks like it." But then Mastandrea started to hedge. "I don't know. I'm not going to try to defend these numbers. I don't know what the service area of a given pharmacy may be. If they serviced a much larger area, as they may be prone to do in West Virginia, it may be it's a little more in line. I don't know where the prescriptions came from."

The committee's investigators weren't going to let Mastandrea off that easy. The company had discovered a second document in the Sav-Rite file that it later turned over to the committee. It was a copy of a 2009 article about the raid on Sav-Rite and Wooley's affiliated drugstore, Sav-Rite #2, at the sham clinic in Stonecoal, four miles up the two-lane highway from Kermit. The *Huntington Herald-Dispatch* reporter who wrote the story now worked as spokesman for Attorney General Patrick Morrisey. Miami-Luken paid $2.5 million to settle the attorney general's lawsuit in 2016, claiming then that the company was poised to close.

"Looking at this article, would you continue doing business with these pharmacies?" Slobodin asked.

"No."

But Mastandrea had more to confess. He had something he wanted to get off his chest. Why else would he have agreed to the interview? Miami-Luken's reputation was at stake. "I am mortified that I have to come and, before this committee, be presented with these figures. It is embarrassing to me that my father's legacy is tainted by this activity. How in God's name we participated in supplying this product when, in hindsight, clearly this was drug diversion. A picture of this pharmacy would be next to the definition in the dictionary. No one was paying attention. It's an abomination."

There, he had said it. His company was as guilty as anyone else. But Miami-Luken and the other distributors didn't act alone, Mastandrea insisted. He reverted to industry talking points—or was he echoing Sergeant Mike Smith's cartel comparison? "Where is the physician oversight that allows this to go on?" Mastandrea wondered aloud. "Where is the state pharmacy board in terms of regulating the practice of pharmacy and letting this continue? The wholesaler has the responsibility to monitor suspicious orders. Where is the responsibility of the manufacturer who, at least at one point in time, promoted Class II narcotics as being an answer to chronic pain and the potential for abuse and addiction being limited? Where is the responsibility of the DEA?"

Mastandrea mentioned that Miami-Luken now had investigators on retainer and a new computer system to monitor its customers. It was the same story. The distributors had been promising to clean up their act since 2007. Their compliance programs didn't work, or maybe nobody wanted them to work. Yet they kept insisting their systems for flagging disreputable pharmacies were infallible.

"How about monitoring news articles that might have your customers named in a raid being part of a pill network?" Slobodin asked.

"Scary. It's scary."

The DEA finally delivered thousands of records to the congressional committee, but blacked out large swaths of text. Nobody could make any sense of it. The DEA redacted entire pages. Greg Walden and a handful of committee members—Diana DeGette of Colorado, Gregg Harper of Mississippi, and David McKinley of West Virginia—gathered at the Capitol for a press conference in early February 2018.

"You know what we got back?" Walden said, flipping through the stack. "We got back a lot of black ink. . . . We've had it with the DEA."

"Uh, this is not helpful," echoed Harper.

While waiting for the DEA's response, Walden's committee had done more investigating on its own. It discovered that distributors had delivered 20 million prescription opioids to two pharmacies in the town of Williamson, West Virginia, population three thousand. The panel asked the companies—the Big Three, plus Miami-Luken and H. D. Smith—for an explanation. It hadn't heard back. The DEA was no help either.

"The DEA simply refuses to cooperate," said DeGette.

She was replaced at the microphone by McKinley, the congressman from West Virginia. He was going last. McKinley recited the state's overdose death rate, which was three times the national average. "You try to understand what's the cause, what's the root cause. Is it economics? Is it education level?"

At a recent town hall, McKinley had fielded phone calls from constituents who complained that not enough people were being arrested. The people poisoning the region weren't being prosecuted. McKinley told them that eight doctors and seven pharmacists had lost their licenses in recent years—information he reiterated at the press conference.

"And we have some number of lawsuits that have caught—about forty-six million to fifty million dollars has been collected from pharmaceuticals," he said.

Congressional aides had also examined shipments to War, a town in southern West Virginia with eight hundred people. "Yet the pharmacy sent eight hundred thousand oxycodone into that community in the last two years." McKinley was careful with his words. Not once did he blame distributors. He closed his remarks by holding up a book about the early days of the opioid crisis. "It was written, in many respects, for my office." McKinley pointed to the front cover. "We use this as a textbook because here it talks about how we should be dealing with it in Washington." The book was written by an expert witness being paid by the drug distributors.

26

Bankrupt

Back at my desk in the capitol pressroom, I was starting to panic. About a week earlier, *Gazette-Mail* publisher Susan Chilton had distributed a "Dear Employee" letter to the staff. "Our dream of an independent newspaper is no longer a financial reality," she wrote. "The Daily Gazette Company and Charleston Newspapers will file a voluntary Chapter 11 bankruptcy petition tomorrow."

We were bankrupt, broke, kaput. That wasn't the worst of it. The *Gazette-Mail* was being sold to Ogden Newspapers, a chain with a reputation for laying off scores of employees, cutting newsroom budgets to the bone, and pandering to public officials and powerful interests. Patrick Morrisey could barely contain himself, exulting on Twitter by quoting the words of Martin Luther King Jr.: "The arc of the moral universe is long, but it bends toward justice." We received our WARN notices next. The alerts were required if a company planned to lay off more than fifty people, and the newspaper employed one hundred and seventy full-time. Ogden, which owned forty-two daily newspapers across sixteen states, was going to make us reapply for our jobs. The company had laid off more than thirty employees, including the entire photography department, at the last paper it gobbled up.

For months, we had been bracing for bankruptcy, ever since a federal judge upheld an arbitrator's decision that the Chilton family had improperly sold the dailymail.com Internet address to the *Daily Mail* of London. The arbitrator and judge ordered the Chiltons to pay $3.8 million to MediaNews Group, the former owners of the *Charleston Daily Mail*. The website address had belonged to MediaNews, not the *Gazette-Mail*. And MediaNews never gave the Chiltons permission to sell it. The $3.8 million tab also included unpaid management fees.

The Daily Gazette Company simply didn't have the cash on hand to settle the dispute. The paper was drowning in debt. It owed the bank $15.6 million on the loan used to buy the *Daily Mail* in 2004, plus another $12 million to the Pension Benefit Guaranty Corporation, a federal agency that guarantees retirement plans, because the paper hadn't paid into its pension plan in years. Ogden made a $10.9 million deal to pick up the pieces after bankruptcy, with a thirty-day window for other companies to bid on the paper. If someone topped Ogden's offer, there would be an auction. Nobody gave that much of a chance. The sale to Ogden seemed a foregone conclusion. Their executives had already made plans to stop by the newsroom in two weeks. Susan did her best to soften the blow, calling the sale "an opportunity to have even more significant statewide reach through other Ogden-owned newspapers in West Virginia." But nobody believed a word. No, this was an embarrassment.

Morrisey wasn't the only one to pile on. The newspaper's critics had always blamed its pro-labor bent and liberal editorials for stifling the state's economy. Ogden, by contrast, was a booster for big business, and its editorials were unabashedly conservative. "The paper's bias against voters who elected President Trump by the widest margin of any state just reached an unbearable level, and it cost them their company," commented Allen Whitt, who headed the conservative Family Policy Council of West Virginia. "This day of reckoning has been coming for a while for that paper." One coal industry lawyer, speaking at a mining conference a day after the bankruptcy announcement, mocked the paper by holding up a fake pink slip made out for Ken Ward, our dogged environmental reporter—eliciting laughter from the audience of energy executives.

Not everyone, however, was celebrating the ownership change. A petition was circulating that asked the bankruptcy judge to reject the sale to Ogden. "The *Gazette* actually covered both sides of issues, not just whatever side money was on," wrote Vivian Stockman in a blog post for the Ohio Valley Environmental Coalition. "The workers, not just the bosses. Environmentalists and people concerned about health, not just the PR spokespeople for industry. Poor people, not just the well-heeled and well-connected. Put simply, West Virginia can't afford this loss."

The *New York Times* also was on the story. In April 2017, my articles about the opioid epidemic were awarded the Pulitzer Prize for investigative reporting, and nine months later, the newspaper was bankrupt. A *Times* reporter called me for comment. I would be reapplying for my job, attending a mandatory résumé-building workshop. "I guess life is full of highs and lows," I told the reporter, "and obviously we're at a low." But "the news goes on, and we're there to cover it."

Things were looking grim. Rob Byers, our executive editor, was sending out staff memos to keep our spirits up, reminding us we were part of a newspaper known across the nation as a "hard charger, an ass-kicker, an advocate for the people of West Virginia and Appalachia." But for how long? We were running out of time. We searched for another buyer. That was the best idea we had, but we were reporters, not brokers. Someone phoned Jay Rockefeller, who had represented West Virginia in the Senate for thirty years. He wasn't interested. Nobody would be clamoring to spend $11 million on a newspaper that just went belly-up, and if they did, they'd have to throw in an extra $500,000 "breakup fee" made payable to Ogden Newspapers.

On a Friday afternoon in February, five of us—me, Byers, Ward, managing editor Greg Moore, and editorial page editor

Dawn Miller—slipped into the newspaper's conference room to prepare for Ogden's first visit to the newsroom. Byers wanted to make sure we were all on the same page and would deliver a uniform message. Ogden CEO Robert Nutting and his daughter would be here at nine o'clock the following Monday morning. They were staying two days, meeting first with Byers and other department heads. The managing editor of Ogden's flagship paper planned to stop by that Monday afternoon and would stick around for meetings with small groups of reporters on Tuesday.

Byers informed us that Nutting had, so far, said all the right things. No, his papers hadn't done a whiff of investigative reporting, but the hard-hitting *Gazette-Mail* could lead by example. Nutting, whose family also owned the Pittsburgh Pirates baseball team, had promised not to mess with our success. But could we believe him? We weren't sure how to act. We weren't going to pretend, play make-believe, and give Ogden a warm welcome. But we didn't want to be rude either. We realized we all might be without a job after the deal closed. Byers and Ward would likely be the first casualties of Ogden's budget ax. And the company wouldn't need an editorial page editor with a progressive voice. My fate was uncertain, too. A decade earlier, I had reported that Ogden had paid to sponsor President George W. Bush's visit to Wheeling, where he gave a speech defending the war in Iraq. A former Ogden editor told me the family hadn't forgotten who wrote the article. I didn't see Greg staying either. There would be a new regime of editors—and fewer reporters. So nobody's job was safe. Why pretend otherwise?

We were skeptical to say the least, so much so that we contemplated something that was sure to throw gasoline on the fire. A New York film producer who had read about the paper's plight was working on a short documentary about us under a contract with the Chiltons. Could his crew film Ogden's visit? It was a terrible idea. It would blindside Nutting and his execs. They'd be livid. It would show a lack of respect. And once they owned the

Gazette-Mail, they'd find out who was responsible and show us the door. But we were willing to take our chances. The film crew would be there first thing Monday morning. Susan Chilton gave them permission to chronicle her last days at the helm of a newspaper her family was giving up after 110 years.

The Nuttings' visit didn't last long. Bob Nutting was in the middle of an important conversation with Byers, gauging his concerns about the transition and any support the newsroom might need, when Nutting spotted the *Gazette-Mail*'s general manager, Trip Shumate, escorting the film crew around the newsroom. The Nuttings promptly packed up their belongings and departed. All meetings scheduled for Tuesday were canceled. We never saw them again.

Two weeks later, word spread that a second bidder was willing to pay $11.5 million plus the breakup fee, and Ogden had no desire to get into a bidding war. The Nuttings withdrew their offer. They would take their money and buy newspapers elsewhere, where they would be appreciated, where employees would treat them with respect, not suspicion. They wished us well. The auction didn't turn up any other interest.

And who emerged as the new owner of the *Gazette-Mail*? It was a wealthy Huntington businessman already familiar with the news industry. He owned the *Huntington Herald-Dispatch* and five other newspapers in southern West Virginia. He had cobbled together a cadre of minority investors, raising just enough money, combined with his own, to pull off the eleventh-hour purchase. Taking a big chance on us was Doug Reynolds, the former state lawmaker and candidate for attorney general in 2016, the Democrat who lost to Patrick Morrisey. The attorney general's Twitter account was silent for once.

27

Paper and Tapes

How did it happen? How did 780 million painkillers spew into West Virginia and nobody said a word? How did millions of pills get dumped in small towns? How did the doctors and the pharmacists and the distributors get away with it?

On a Tuesday morning in mid-March of 2018, the congressional panel called on DEA acting administrator Robert Patterson, the agency's highest-ranking officer, to testify about what went wrong in West Virginia. By then, the DEA had reluctantly turned over more of its pill shipment data to Congress. The numbers raised even more questions about the DEA's inaction. Gregg Harper, who was chairman of the House Energy Committee's Subcommittee on Oversight and Investigations, recognized himself to go first.

"Would you agree on its face these distribution figures represent inordinate amounts of opioids shipped to rural markets?"

"I would," Patterson said.

For as long as anyone could remember, the DEA had tracked painkillers and other controlled substances with a computer program called the Automation of Reports and Consolidated Orders System, or ARCOS. It monitored sales from drug manufacturers to distributors to pharmacies and hospitals. It didn't miss anything.

But somebody would have to check it. It didn't send automatic alerts.

"Ten years ago, would the ARCOS database have been able to flag DEA diversion investigators about stunning monthly increases of shipment amounts?" Harper asked, reading from a sheet of prepared questions.

"Ten years ago, I think that would be doubtful. Sir, so ARCOS data probably pre-2010 was an extremely manual process."

What was Patterson saying? That the DEA was counting pill sales numbers by hand? Surely not. It was an automated system. The DEA had computers back then, too.

Congresswoman Diana DeGette wanted a clarification. "And so what was happening? The data was just being reported, but nothing was being done with it? Isn't that correct?"

"I would say it was used in a very reactive way," Patterson said, twirling a pen.

DeGette referred the DEA chief to a letter the committee had sent to McKesson's CEO, John Hammergren. The ARCOS data showed that McKesson had shipped 2.2 million hydrocodone and 78,500 oxycodone pills to Sav-Rite Pharmacy in Kermit in 2006, and another 2.6 million hydrocodone and 40,900 oxycodone tablets the following year. And that was just McKesson. Sav-Rite purchased painkillers from other distributors those years as well. "I'm going to assume," DeGette said, "if we had been analyzing this data, we would have found the 184,000 pills per month McKesson was sending to Kermit, if someone had looked at it? Wouldn't you think so?"

"I do agree with that."

Federal law required distributors to "know their customers." Just like the DEA, distributors were responsible for identifying suspect orders.

"The key burden is actually on the distributor," Patterson said.

"Exactly," DeGette said. The DEA chief had teed up her next question. "So do you think if you were McKesson Corporation, and you were looking at all the prescriptions in Kermit, that you would think that, would you think they knew those customers?"

"I think McKesson's answer would be that, you know, they did their part on this."

"Well, what's your answer?"

"Obviously, I think they should have done more," Patterson said.

"Do you think that the distributors failed to adequately exercise good due diligence over what they were doing?"

"I can't tell you what their due diligence was, but—"

"Oh, we're going to ask them that," DeGette said. "Don't worry."

For now, only the DEA had come to answer questions, and committee members were determined to find out why nothing was done while the agency had the damning ARCOS data staring it in the face.

"Did you have any kind of data analysts?" asked Congressman Frank Pallone of New Jersey.

"So my understanding of the people that were handling the ARCOS data, it was a completely manual process, meaning everything was coming in on paper or tapes." Paper. Tapes. Patterson was putting the blame on the reporting system, not DEA personnel. The technology just wasn't good enough. "So you have this one-month to three-month delay to begin with, and they would have errors in their report. So what you found yourself with was a set of data that would sometimes take a year-plus to correct."

Patterson assured Pallone that the ARCOS reporting system had been reprogrammed to catch monthly and yearly surges in pain pill shipments. What happened in Kermit would never again happen. That he could guarantee. But committee members weren't going to let the DEA off the hook with a would-have, could-have, should-have. Congressman Raul Ruiz, a Democrat from California, delivered a parting shot. Ruiz was a doctor who specialized in emergency medicine. He understood what it's like to have to resuscitate people who've overdosed. They would be dumped off at the emergency room, unconscious, blue in the face, not breathing. Some would live. Some would die. He didn't want to hear any excuses from the DEA. It was too late for excuses.

"You screwed up," Ruiz told Patterson. "You were collecting data you didn't know how to use. You weren't paying attention to your job."

Patterson, who had worked at the DEA for thirty years, stepped down three months later. He emailed staff that managing the agency as a temporary fill-in had become "increasingly challenging."

The same week Patterson testified, Patrick Morrisey logged on to his Twitter account on a Sunday evening to assail a forty-three-year-old hospital security guard. Morrisey was running for US Senate in the Republican primary, the election was a month away, and this clown, this rent-a-cop with forty-seven measly followers, had the audacity to yap about Morrisey's ties to Big Pharma.

Morrisey was notorious for his late-night Twitter spats, and nothing got under his skin more than when critics brought up his lobbying history. But the security guard, Todd Wentz, had started it. In a tweet, Morrisey had attacked his GOP primary opponent, Congressman Evan Jenkins, over gun control, and Wentz responded, "How many pain pills did he flood our state with?" Morrisey punched back: "Would you believe none? And dramatically reduced overall totals more than anyone else? Amazing how the truth sounds! No one has done more to take on excess supply."

Wentz's next response brought up Morrisey's wife and Cardinal Health. "So your wife isn't a lobbyist for big pharma? Your firm doesn't handle cardinal health's business. And then why has the media named you 'pain pill pat?'" Wentz summed up with a blunt, albeit misspelled, epithet: "Everyone knows you're a schill for @cardinalhealth." Morrisey lashed back, "It's all false and you are a troll!"

I became aware of the testy exchange in an article by PolitiFact, an organization that analyzes the veracity of candidates' statements in select races across the country. The article cited Morrisey's lobbying for the Healthcare Distribution Management Association (now called the Healthcare Distribution Alliance), a trade group that represents wholesale drug distributors. PolitiFact also spotlighted Denise Henry Morrisey's lobbying for Cardinal, but the article noted, "We were unable to confirm whether any of Denise Henry Morrisey's lobbying work addressed opioids." Her DC lobbying firm, Capitol Counsel, ignored PolitiFact's request for comment.

Back in the fall of 2016, I had reported Cardinal's statement that Denise Henry Morrisey had never lobbied on prescription drugs. Days later, I did a cursory check. In fact, she did represent Cardinal on prescription drug issues. I sent the company's spokeswoman copies of Denise Henry Morrisey's disclosure forms that she personally filed with Congress. The forms left no doubt. "You were right. We were wrong," the spokeswoman told me over the phone. I had prepared for a stubborn denial. Instead, I got an admission. It was almost unheard of for a corporation, especially one the size of Cardinal, to admit they'd released a false statement. Maybe they'd acknowledge they "misspoke." But never "We were wrong." I decided their mea culpa didn't deserve a follow-up story—at least not right away. Turns out, I would wait a year and a half.

After PolitiFact published its article, I combed through even more of Denise Henry Morrisey's lobbying disclosures. While being paid by Cardinal, she had lobbied federal lawmakers on opioid issues, specifically against legislation and rules that aimed to impose tighter restrictions on hydrocodone. She did so while her husband's office was overseeing the lawsuit that accused Cardinal of shipping an excessive number of hydrocodone pills to West Virginia. The legislation, which eventually passed, made it harder to prescribe hydrocodone. The reclassification led to a sharp decrease in the number of hydrocodone pills dispensed in

West Virginia and other states. Hydrocodone-related overdose deaths plunged.

More evidence showed that Cardinal's go-to lobbyist had sought to influence federal lawmakers on prescription opioids. Earlier in the year, a source had forwarded a copy of an email Denise Henry Morrisey had sent to West Virginia congressman David McKinley's office in late February 2012. The request was designed to take Cardinal off the hot seat and put the DEA on it. That day, her client lost a court battle with the DEA and was ordered to stop shipping pain pills from one of its busiest warehouses. A House of Representatives hearing on the nation's prescription drug scourge was scheduled for the following morning, and Denise Morrisey had a list of questions she wanted McKinley to ask.

A month earlier, the DEA had abruptly suspended Cardinal's license to distribute OxyContin from its warehouse in central Florida owing to "staggeringly high" and "exponentially increasing" shipments to notorious pill mill pharmacies that posed an "imminent danger to the public health and safety." The company went to court to fight the order, arguing that the suspension would tarnish its "business reputation" and scare off customers. Cardinal got in trouble with the DEA before, paying a $34 million fine in 2008 for failing to block suspicious orders for painkillers from four warehouses, including the Florida distribution center that the DEA was now targeting. Cardinal had signed an agreement with the DEA pledging not to allow the same thing to happen again. But it did. For that reason, on the eve of the House hearing a federal judge rejected Cardinal's effort to block the DEA's order to shut down the warehouse.

Given Cardinal's history with the DEA, Denise Morrisey had reason to expect distributors would come in for blame under testimony from one witness in particular: Joe Rannazzisi, the agency's top enforcer. She sent McKinley's top aides an email with a series of questions and statements that he could use to grill Rannazzisi, who oversaw the DEA's anti-drug-diversion program. Cardinal had furnished McKinley—and presumably other subcommittee members—with a script. All they had to do was read it.

"Please help us understand . . . why should a distributor be punished by the DEA for distributing drugs to a pharmacy that has an active DEA registration?" Denise Morrisey asked McKinley to ask Rannazzisi. Another scripted question included Rannazzisi's name: "Please explain to the subcommittee, Mr. Rannazzisi, doesn't an overemphasis on enforcement with distributors create an impossible know-your-customers standard in law enforcement?" And there was this blame shifter: "After all, at the end of the day, the abuse we are talking about relates to consumer behavior at the bottom of the chain. Basically, we're talking about drug addicts getting prescriptions from questionable doctors and filling them at pharmacies which are either looking the other way or not doing enough to be vigilant."

I searched YouTube and found a video recording of the 2012 congressional hearing. C-SPAN had covered it. I watched and took notes. Denise Morrisey was sitting in the second row. She had on a black dress. She was typing on an iPad.

After the hearing started, Joe Rannazzisi used his opening statement to describe the epidemic from the front lines. He seemed to be unaware that subcommittee members had been "teed up"—as lobbyists call the practice of preparing lawmakers—to rattle him.

"They are not valid prescriptions," Rannazzisi said, "and the wholesalers and distributors just continue to ship large amounts of drugs to those pharmacies without doing due diligence, without knowing their customer, without saying, 'Well, why are you ordering? Why are you ordering this amount of drug when every other average pharmacy in the US orders this, and you are ten, twelve, fourteen times more than that?'"

Then Congressman Gregg Harper of Mississippi read off questions from the email Denise Morrisey had sent McKinley.

"What regulation does DEA have that specifically outlined

the legal requirements that pharmacies and distributors and manufacturers are required to take to avoid drug diversion?" Harper asked, repeating verbatim from the email.

Rannazzisi cited regulations. One required pharmacists to make sure prescriptions are valid. Others mandated that manufacturers and distributors monitor and report suspicious orders for huge numbers of drugs.

"OK," Harper said.

The Mississippi congressman had more questions: Did the DEA pass along thresholds to distributors that they could use to identify questionable orders? Why doesn't the DEA share with distributors the number of pills their competitors are shipping to the same pharmacies? He read again from Denise Morrisey's script:

"What guidance has DEA provided to the manufacturers, distributors, pharmacies, or whatever on the specific steps that they should be taking to identify fraudulent prescriptions?" Harper added the "or whatever."

"Well, there are certain red flags," Rannazzisi explained. "For instance, a pharmacy. If you have, if you are sitting in, we will say Portsmouth, Ohio, and all of your customers are coming from, I don't know, eighty or a hundred miles away, and the doctor you are filling for is a hundred miles the opposite way, and it is all cash transactions, and you are seeing this over and over again, you know, I am not the smartest guy, but red flags pop up in my mind when that happens."

Then McKinley got his shot. Having advocated for a national prescription drug registry during his opening statement, he now wanted to know why state computer systems that tracked doctors' prescriptions couldn't be connected. The attorneys general testifying alongside Rannazzisi told McKinley that was a work in progress.

The next panel represented the drug industry, and the subcommittee members had only softballs to throw them.

McKinley asked John Gray, executive director of the Healthcare Distribution Management Association, about its meetings with DEA officials—a line of inquiry Denise Morrisey suggested in her email the day before. "Are the distributors getting good advice, good direction when they go to the DEA and ask for improvements to their delivery system before they pull the registration?"

Gray characterized the meetings as "deficient." The DEA wouldn't tell distributors when to stop selling pills to pharmacies. The distributors had spent "tens of millions of dollars" setting up systems to flag suspicious orders, Gray said. McKinley took off his wire-rim glasses and rubbed his eyes. This cost to the companies was an unfunded mandate, the congressman said.

"I have this nagging feeling here that there are parts of the chain that are not being treated equally, and I hope that the DEA will revisit how they work with each," McKinley said.

"Well, we do, too," Gray said.

Six years later, while running in the GOP primary for the US Senate, Patrick Morrisey was sidestepping our questions about his wife's lobbying. "It's wrong that candidates for office attack my wife when she has nothing to do with opiate issues in West Virginia. Nothing," he told our political reporter, Jake Zuckerman, during a stop by President Donald Trump in West Virginia. "My wife never worked on any West Virginia issues." Jake repeated the question. Morrisey said, "I've never worked on opiate issues in the private sector." Never mind that Jake had asked about Denise Morrisey's lobbying work, not his.

Morrisey shrugged off the criticism, played up his settlements with drug companies, and won the primary. Now, only one person stood in his way for a seat in the US Senate—incumbent senator Joe Manchin. Manchin was a Democrat in a deep-red state. And Morrisey had the backing of the president. Trump would make three visits to West Virginia to campaign for Morrisey. They shared a propensity for stirring up trouble on Twitter.

As for Wentz, PolitiFact reported that the security guard was "a little shocked" that the attorney general "went at me like that" on Twitter. "Really, the only forum an ordinary Joe like me has to voice my opinion straight to him is by social media," Wentz said. A registered Republican, Wentz disclosed he was a recovering opioid addict. "So I know the perils from that end, and now that I've turned my life around and work security in a hospital, I see the desperation in these people's eyes." Wentz, "through the grace of God," hadn't used in six years.

28

"What's the Punishment?"

Top executives from five of the largest drug distributors in America stood side by side, raised their right hands, and swore to tell the truth, the whole truth, and nothing but. They were on Capitol Hill to testify about the opioid crisis. It was May 8, 2018—exactly one year after Congress started its investigation.

Hearing Room 2123 at the Rayburn House Office Building was packed. Photographers crouched below the dais. After being sworn in, the five witnesses—McKesson CEO John Hammergren, Cardinal Health chairman George Barrett, Amerisource-Bergen CEO Steven Collis, Miami-Luken chairman Joseph Mastandrea, and H. D. Smith past president J. Chris Smith—took seats at a conference table.

The night before, I had scrambled to write a preview story based on Barrett's prepared remarks, which had been posted on the House of Representatives' website. Barrett apologized for the company's role in failing to stop the flow of highly addictive painkillers, the first time anyone with the Big Three expressed even a hint of remorse. "To the people of West Virginia, I want to express my personal regret for judgments that we'd make differently today with regard to two pharmacies that have been a particular focus of this subcommittee," Barrett said in the prepared statement. "With the benefit of hindsight, I wished we had moved faster and asked a different set of questions. I'm deeply sorry we did not."

The following morning, after Barrett—the only executive to issue an apology—and his four colleagues delivered their opening remarks, the subcommittee's chairman, Gregg Harper, wasted no time with niceties. Six years earlier, at a similar hearing, Harper had brought questions supplied by Cardinal lobbyist Denise Henry Morrisey. Not this time. He directed his opening salvo to all the distributors, starting with Barrett:

"Do you believe that the actions that you or your company took contributed to the opioid epidemic?"

"Thank you, Mr. Chairman—"

Harper stopped Barrett. "We're really looking here, because I've got a lot of questions, for yes or no."

"No. No, sir, I do not believe that we contributed to the opioid crisis." An apology, yes. An admission of guilt, no.

"Dr. Mastandrea?"

"Yes."

"Mr. Hammergren?"

"No."

"Mr. Smith?"

"I believe H. D. Smith conducted itself responsibly and discharged its obligations."

"Is that a no?" Harper asked.

"That is a no."

"Mr. Collis?"

"No. I believe we—" Collis paused. "It's a no for Amerisource-Bergen."

Harper's questioning, going down the row, one by one, stoked memories of a 1994 congressional hearing in the same building, where seven tobacco executives testified that nicotine wasn't addictive. The distributors, however, weren't unanimous. Mastandrea's admission broke ranks. His company had less to lose. It was reportedly going out of business.

Harper had a second yes-or-no question, but as in the first go-around, the distributors didn't want to be held to one-word answers. Their lawyers sat behind them in the front row. "Do you acknowledge that your company had past failings in maintaining effective controls to prevent the diversion of opioids?" Harper called on the executives in the same order.

"I have no reason to challenge the good faith of the decisions made by people many years ago," Barrett said. "But I can say that the decisions we might have made on some of those pharmacies would look differently today."

The subcommittee had lambasted Cardinal for shipping 6.5 million prescription opioids in four years to Family Discount Pharmacy, the small-town store in rural Logan County, which had the thirteenth-highest overdose death rate in America. On top of that, McKesson and H. D. Smith had supplied the same pharmacy with millions of painkillers. Cardinal had also sold another 1.2 million pain pills over two years to a drugstore that filled prescriptions for the infamous Wellness Center in Williamson, until the Feds shut it down.

"Is that a no?" Harper asked Barrett.

"I think our organization understood its obligations."

"So is that a yes that's now a no? I'm a little—"

"I am looking back on history. And what I'm describing is an organization that I believe did its job at the time."

Harper had to move on. His five minutes was winding down. "Dr. Mastandrea, do you acknowledge your company had past failings to prevent the diversion of opioids?"

"Yes."

Hammergren was next. "Our organization has worked for decades to try to meet our obligations under the DEA regulations."

"It seems like a pretty simple question," Harper said.

"In the past we've had challenges understanding the expectations that our regulator would like us to follow."

"Mr. Smith?"

"Again, I believe H. D. Smith has acted responsibly. So the answer would be no."

Collis had the advantage of going last. "I believe we've always discharged our duties effectively and responsibly and have maintained an adequate diversion program."

The panel had asked the distributors to turn over everything in their files about Sav-Rite Pharmacy. Miami-Luken, the records showed, shipped to the Kermit drugstore even after the DEA raided it. McKesson did the same, but after ceasing shipments for two years. Congresswoman Diana DeGette reminded the executives that the DEA had sent them a letter in 2007, outlining their responsibilities to screen pharmacies that bought their drugs. "Now, Dr. Mastandrea, we asked Miami-Luken to provide us with its entire due diligence file on the Sav-Rite Pharmacy, and this is what we got from you." DeGette waved a thin stack of papers. "Do you recognize these documents?"

"No."

"We can have somebody hand them to you, but I will assure you it's about fifteen pages of purchase orders and sales orders. Do you think this is a sufficient due-diligence file for all of the number of opioids that you were sending to Kermit, West Virginia?"

"No."

"Thank you. And you know what? Thank you for your honesty today."

McKesson had turned over even less about Sav-Rite—two pages. The company had shipped more than 5.6 million pain pills to the pharmacy. "Do you recognize this document, sir?" DeGette asked, showing it to Hammergren. The form, signed by Jim Wooley, was submitted to McKesson in 2007, declaring that Sav-Rite filled only legitimate prescriptions.

"No, I don't." The document was marked Exhibit 3 in a three-ring binder provided to each executive. The exhibits had different-colored tabs. "This is the first time I've seen this document."

"This is the only documentation that McKesson gave to this

committee when asked for the due diligence file for Sav-Rite," DeGette said.

"I believe our relationship should have been terminated with Sav-Rite immediately."

"Yes or no, do you think this is sufficient documentation to show compliance with the rules of the DEA?"

"We continue to evolve our diligence—"

"Yes or no will work, sir."

"I've not reviewed the document. I can't provide an answer to that."

Congressman Greg Walden had reviewed the documents. He found a page stating that McKesson policy limited Sav-Rite to eight thousand prescription opioids per month—part of the program started in 2007 to monitor high-volume orders. Yet McKesson sold 289,500 hydrocodone pills a month to the Kermit pharmacy.

"Your own distribution was thirty-six times higher than the threshold you set," Walden said. "So did this program identify Sav-Rite?"

"It did not, sir. Our systems at the time weren't automated enough, certainly, and we didn't flag it fast enough."

H. D. Smith's shipments also drew scrutiny. Chris Smith, who'd stepped down as president after selling the company to AmerisourceBergen, had earlier testified that it always operated on the straight and narrow. But Congresswoman Susan Brooks, a former federal prosecutor, had copies of internal emails that suggested otherwise—management knew a lot more than it was letting on. In 2016, an H. D. Smith employee tipped off executives to the improper vetting of a new pharmacy in Logan County, allowing it to exceed its monthly limit for hydrocodone in twelve days and still qualify for a special monthly discount.

"What monthly discount?" Brooks asked.

"I'm not sure what that refers to," Smith said.

"Monthly discount? No idea what monthly—what deals are being cut?"

"I'm not sure what that refers to."

H. D. Smith stopped shipping to the pharmacy in February 2018, two weeks after I reported that the House panel had expanded its investigation to include the regional distributor. Smith also stepped down that month. "Well, I will say, according to a document we received," Brooks said, "the company cited its reason for taking this action and finally terminating the relationship was due to negative news coverage."

Congressman David McKinley wasn't a member of the subcommittee, but Harper was going to let him speak anyway. McKinley had waited more than an hour, and it was killing him to stay silent. This was about his state, West Virginia. In the past, he had avoided placing blame on distributors. But that was then. The companies and their lobbyists would have to find someone else to do their bidding. A thousand West Virginians had overdosed and died the previous year.

"The fury inside me is bubbling over," McKinley said. He had heard enough about the distributors' fancy algorithms, their infallible systems for catching suspicious orders, and their "we only fill orders for licensed pharmacies" explanations. Weren't the CEOs responsible for setting the tone at the top? "And for several of you to say you had no role whatsoever in this—no, you had a role," McKinley said. "You had a role."

McKinley stared down Hammergren, chastising his company for failing to send suspicious-order reports to the Board of Pharmacy before 2012. "That was the genesis. That's when the disease really took hold in West Virginia. But you weren't complying. Yet you said, 'We're not responsible.' I think you very much were responsible."

Hammergren waited for McKinley to finish, but the congressman was just getting started. Should doctors who write illegal prescriptions be held accountable? Yes, Hammergren said. Should rogue pharmacists such as Sav-Rite's owner be held accountable? Yes, Hammergren said. McKinley was trying to set something up, to make a point, to assign responsibility. "Do you regret any role that your company has played in this crisis?"

"Congressman, I don't know how you could look at this crisis and not feel terrible about what's going on in this country."

"So you do regret?"

"I feel terrible about this crisis."

"What's the punishment?" McKinley asked. A fine equivalent to one-tenth of 1 percent of yearly revenue? A slap on the wrist? Were settlement agreements just a cost of doing business? Were financial penalties enough to send a message? "Or should there be time spent for participating in this?" McKinley asked, now directing the question to all five men at the table below. Nobody said a word. By "time spent" was McKinley saying they should be charged with crimes and sent to prison? "I just want you to feel shame about your roles, respectively, in all this."

As the hearing drew to a close, Harper asked each CEO whether his company would keep watch for excessive orders of prescription painkillers, not just from pharmacies in West Virginia, but in small towns across America.

"Sir, we will and we do," Barrett said.

"Absolutely," said Mastandrea.

"Absolutely," Hammergren repeated.

"I'm not in a position to do that," Smith said.

"We will," Collis said. "Unfortunately, opioids seem to thrive in communities where there often is, you know, hardship. And so we feel particularly concerned about that."

The investigation stopped then and there. There were no more hearings about pill dumping in West Virginia, no more witnesses. A report by Politico offered a possible explanation for the abrupt ending. Through their political action committees, AmerisourceBergen, McKesson, and Cardinal had contributed a combined $180,000 to the campaigns of House Energy and Commerce Committee members during the most recent election cycle. With an election in November—and Republican control of the House of Representatives at stake—there'd be plenty more where that came from.

Five days before Christmas 2018, after Democrats won control of the House of Representatives in the midterm elections, the Energy and Commerce Committee released a 324-page report that excoriated the prescription drug distributors. McKesson, Cardinal Health, AmerisourceBergen, H. D. Smith, Miami-Luken—the report skewered each and every one of them. Time and again, there were "breakdowns." Distributors set limits on pharmacy drug orders but failed to enforce them. They flagged excessive shipments but didn't stop them. They sold drugs to pharmacies they knew were breaking the law. They "willfully ignored" red flags and stark warning signs that the drugs they sold were being diverted to the black market. When one distributor cut off a pharmacy, another supplier stepped in. The excuse that the distributors didn't know what their competitors were selling to the same pharmacy was nonsense. They all could request pharmacy dispensing reports. They could see the enormous numbers of painkillers going out of every drugstore door. The "missteps" and "missed opportunities" contributed to the worsening of the opioid crisis, the report concluded.

"I don't buy this staggering level of ignorance at all," a Republican staff member who investigated the companies for Walden's committee told me. "They knew a lot but just didn't do anything. It was just sell, sell, sell, and, hey, we'll just look the other way."

The distributors' high-paid lawyers, their arrogance, their smugness, it was hard to take. The massive shipments, the suspect orders? The company lawyers wondered what all the fuss was about. Why were they being harassed?

"They acted like we were the ones who were crazy," Walden's staffer said.

The report also blasted the DEA. The agency didn't use its database to monitor the flood of prescription painkillers to West Virginia. It backed off enforcement. It inexplicably missed that millions of pills were being sent to small towns. Infighting within the DEA was a distraction.

The report reiterated that the DEA put the distributors on high alert about their legal duty to report and block suspicious drug orders. In 2005 and 2006, DEA administrators met with the companies' compliance officers—in some cases, two or three times. The companies received letters about their obligations to stop pills from being diverted. They stopped nothing. Instead, they opened the floodgates, paid fines, then looked the other way.

In response to the report, distributors seized on a single sentence in the executive summary that alluded to how doctors, pharmacists, manufacturers, and drug traffickers also played a part in starting and fueling the opioid crisis.

"Cardinal Health commends the House Energy and Commerce Committee for, as it said, 'a look at a piece of the overall puzzle' of this complex national public health issue," the company stated in a press release. The wholesale distributors were still spouting stubborn denials, while, remarkably, praising the report as if it had cleared them of any wrongdoing.

The report ominously mentioned that drug overdose deaths were driving a decline in life expectancy in the United States, while such statistics were improving just about everywhere else in the world. As a parting gift, Republican committee staff sent me a link to nine-hundred-plus pages of documents—many marked CONFIDENTIAL—that the companies had turned over during the congressional investigation. The distributors had asked that the records be kept secret. The evidence was damning, overwhelming, heartbreaking.

"The lives that are lost are not coming back," the staffer said. We were talking about regional distributor Miami-Luken, but he seemed to aim his comments at all distributors. "Lives are not coming back from the way they operated, from the way they did business."

29

Whose Pills?

In the latter half of 2018 and throughout 2019, the *Gazette-Mail* was back in court trying to shake loose more pain pill shipment numbers, only this time, not just for West Virginia, but for the entire country. Cleveland federal court judge Dan Polster was overseeing an ever-expanding number of lawsuits, some fifteen hundred filed by towns, cities, counties, states, labor unions, and Native American tribes. It had started in Kermit, but now Philadelphia and Chicago and New York City were suing the distributors, and once again the companies were fighting to shield their painkiller sales. They had a new ally—the Drug Enforcement Administration.

An email from a former *Gazette-Mail* colleague had grabbed my attention in the spring of 2018. Elaina Sauber, who had worked at the paper when we uncovered the pill shipments to West Virginia, was now a reporter at the *Tennessean* in Nashville, covering local lawsuits filed against the distributors. The cases were being consolidated with all the others in federal court. Eastern Tennessee's overdose death rate rivaled West Virginia's. Judge Polster had just ordered the DEA to release painkiller sales data for all fifty states to lawyers representing state and local governments.

"Our management in Tennessee really doesn't understand how important this data is and that we should be fighting for it to be unsealed," Elaina wrote. "So I've been told to lead the effort in requesting that data once the DEA turns it over at the end of the month."

Elaina's idea was brilliant. If the DEA wouldn't release the nationwide pill numbers—and the agents had told me as much during the 2017 meeting at the Marriott in Charleston—why not get the records that the DEA had turned over to the lawyers representing the cities and counties? All they would have to do was ask their lawyers for copies. The politicians would certainly want the pill numbers to get out. It would shame the drug companies, bolster the lawsuits, maybe spur settlements like what happened with Cardinal and AmerisourceBergen in West Virginia. We had a public records request that local governments would be dying to fulfill, and wouldn't that be a sight to see? It was too easy. Indeed.

The distributors' well-compensated lawyers had anticipated this. They had a court-ordered agreement with the plaintiffs' lawyers that the pill data they reported to the DEA wouldn't be released to the public. The judge had directed the lawyers to alert him immediately if anyone requested the opioid pill numbers. I phoned Pat McGinley. He was eager to get started. There was a public health crisis. That far outweighed the DEA's and distributors' interests in secrecy.

The *Gazette-Mail* and the *Herald-Dispatch* filed a joint public records request with the Cabell County Commission. Cabell County had the highest overdose fatality rate in the nation in 2017, and its lawsuit was at the front of the line for a jury trial. The commission's attorney, Paul Farrell Jr., also was a lead lawyer for all the cases. He had the DEA data for the entire nation, information about opioid shipments to nearly sixty-two thousand retail pharmacies from 2006 to 2014 that the companies had provided. It would reveal other parts of the country that were flooded with pills. Did what had happened in West Virginia happen in eastern Kentucky, southwest Ohio, the Ozarks of Missouri? With the data, we could finally find out.

The newspapers' request—we were the first to ask—triggered Polster's alert. The drug distributors objected and joined forces with the DEA. The distributors insisted their sales data be kept confidential; why else was there a court order? The DEA argued that releasing the pain pill numbers would hamper its investigations and invite criminals to steal drugs from high-volume pharmacies. The agency's lawyers also warned that disclosing the sales data might lead to "press stories." Careless reporters would misinterpret the numbers and unfairly target law-abiding pharmacies. Or worse, the media might interfere with investigations, tipping off pharmacists with questions about their deliveries and sales.

The DEA hid its other objections, blacking out entire sections of its legal brief. It didn't want the public to know about its policies and procedures for tracking sales of addictive drugs. Lawyers for the *Washington Post*, which requested the same records from two Ohio counties days after we asked Cabell County, called out the DEA. The *Post*'s lawyers informed the judge it would be difficult to argue against the DEA's objections, given nobody could see them. The DEA refiled the same brief with fewer sections blacked out, but on the same day our newspaper and the *Post* both had a court-ordered deadline to declare reasons for releasing the nationwide pill numbers. That left no time for our side to read and react. Nonetheless, we had compelling arguments for disclosure. "Lives have been destroyed," wrote the *Post*'s lawyer Karen Lefton. "It's past time for secrecy."

The Cabell County commissioners couldn't have agreed more. They were suing the distributors—and drug manufacturers now, as well—on behalf of the county's citizens. Every city and town was suing on behalf of taxpayers. Why should they be kept in the dark? Whispers were that a "global" settlement could top $50 billion. The Cabell commissioners were unequivocal. They wanted to release the pill numbers if the judge would allow it. "I don't want to withhold any information from the public," said Bob Bailey, the commission's president. "They are the people we work for."

But Polster didn't work for the commissioners, and he would have the final say. The judge's comments from a hearing earlier in the year were cause for optimism that he might lift the lid: "There are certain areas of the country where there are hundreds and thousands of pills per person, per year, for every man, woman, and child. Everyone knows that was wrong. It shouldn't have happened. The question is, whose pills?"

But the judge wasn't going to reveal the answer anytime soon. He scheduled no hearings. He didn't allow McGinley and Lefton to make their arguments in open court. Finally, on July 26, 2018, Polster sided with secrecy. He concluded the painkiller numbers were "sensitive to pharmacies and distributors because it is confidential business information, and it is sensitive from the DEA's perspective because it is crucial to law enforcement efforts." His decision blocked the counties and their lawyers from ever releasing the DEA data—even after the lawsuits were settled or decided by a jury. The distributors couldn't have walked away with a better ruling if they had written it themselves. Yes, the West Virginia numbers got out, and that was unfortunate, and, well, embarrassing, but the shipments to other states would remain confidential. The distributors could breathe easy.

The morning after Polster ruled against us, Al Emch, AmerisourceBergen's lawyer, was on the phone. He was being gracious in victory, passing along some journalistic advice. Eric, he said, you should be going after prescription data from the state Board of Pharmacy. Its database was far more interesting than the DEA's. I had requested the pharmacy board's data years ago, but state law expressly prohibited the release of the prescription information to anyone except law enforcement. I thanked Emch for the suggestion, and then it hit me. This was a dead end anyway. The pharmacy board database tracked doctors' prescriptions and drugstores' dispensing, but it didn't provide a shred of information about distributors. That's where our conversation ended.

McGinley, meanwhile, was already talking with Lefton about

an appeal. Lefton had discovered that dozens of pleadings in the consolidated cases had been filed under seal. And the judge was going along with it. This wasn't a private dispute being litigated in public. It was a public dispute being wrongly argued in private. Didn't the families of the seventy thousand Americans dying each year of drug overdoses deserve to know what started the public health crisis? This was the epidemic of epidemics.

The *Gazette-Mail* and *Herald-Dispatch*, along with the *Washington Post*, appealed Polster's ruling to the US Sixth Circuit Court of Appeals. The same court had issued a landmark First Amendment decision in 2002, siding with the *Detroit Free Press*, which had fought to keep immigration hearings open after the September 11 terrorist attacks. The court's opinion was unanimous. It declared, "Democracies die behind closed doors."

30

Three Feet Deep

It had been ten months since Timmy Dale's funeral parade ended at the Preece family burial plot, yet the only identifier on his grave was a tiny laminated sign—it had his name, picture, date of birth, and date of death—staked in the dirt. Debbie was saving up to buy a custom-engraved granite headstone. She wanted an image of a fire truck etched on it. She wasn't going to settle for some cheap, flat marker.

I had accompanied Debbie to the ridgetop cemetery, her family's section cordoned off by a cinder-block wall that cascaded down the slope toward the river's edge. This early in the evening, the sun slanted through a canopy of oak and hickory and poplar. Debbie showed me the gravesites of her great-grandparents Wilbur and Emma, her grandparents Thomas and Estella, and her parents, Wig and Cooney. Someone had placed a blue-painted rock at the base of Bull's headstone. Another brother, James "Red Ed" Preece, nicknamed on account of his red hair, was buried along the opposite wall. Debbie wondered if he, too, might have died of an overdose. There hadn't been an autopsy. Timmy Dale's grave was still covered by dirt, though weeds had begun to poke through. Grass carpeted the rest of the Preece plot. The past was descending on her again.

"I guess when it hits you at home, you become passionate."

Debbie stared at the ground. "It's not just hit me once. It's potentially three times, you know? And I guess we're paying for what we did, for what my family did."

That was three decades ago. It couldn't be changed. All these years, she had been trying to make up for it. But history was stubborn. Some townsfolk didn't forget. They still branded her family outlaws. They didn't know she had switched sides.

"I've been mad since 2005," she said, her arms crossed. "I'm still mad. Somebody's got to be made aware of what's going on here. These people are making millions of dollars, and they don't care who dies. They flat out don't care."

Debbie just wanted somebody to listen. Not many were listening before 2005. Now, her lawsuit over her brother Bull's fatal overdose at Mud Lick had sprung a tidal wave of cases, and everybody was paying attention. But the individuals responsible—the Dr. Kisers, the Jim Wooleys, the corporate CEOs—had they paid a sufficient price? I remarked that Kiser and Wooley served time in prison. "That don't bother them," Debbie said. "I've been to Club Fed. It's not all that."

I asked her what she thought of the $47 million in settlements with drug distributors, and the potential for billions of dollars more. She didn't trust that the money would go where it's needed. The state settlement didn't lead to a single addiction treatment center being built in southern West Virginia. More populous counties sucked up most of the money. That's how it always went.

The state didn't deliver overdose-reversing naloxone to the Kermit Volunteer Fire Department until after I wrote an article about a shortage of the lifesaving drug in rural counties, and then Tomahawk's cell phone was ringing and the drug was on the way to the firehouse, and the politicians were fighting to take credit. "When I told them the newspaper was here, everything rolled so quickly," Tomahawk had told me at the firehouse earlier in the year. "It just happened so easy and so all of a sudden. Funny how that works."

Debbie hadn't paid much attention to the congressional hearings that took the DEA and the distributors to task. She didn't trust politicians—not in Mingo County or Charleston or Washington. She wasn't even following the Town of Kermit's lawsuit. She still had recovering addicts coming to her home, some wanting to sign on to Cagle's lawsuit against pharmacists and former doctors. The case, after years of delays, was scheduled to go to trial in March 2020, but a settlement was more likely. Other addicts would come to Debbie's house for a meal or a couch to sleep on at night. Her chemo treatments had stopped the spread of the cancer she had been diagnosed with eight years earlier. She had more good days than bad. But still, she needed a part-time home health worker to help around the house, and she was taking multiple medications prescribed by doctors at Duke. The employees at Kermit's only pharmacy—with a new name, a clean slate, and Jim Wooley long gone, though some of his loyal workers had stayed on with the new owners—questioned her prescriptions written by out-of-state oncologists and neurologists. The drugstore filled some scripts, rejected others. Debbie figured they were still sore over Bull's lawsuit and what snowballed afterward.

"I don't normally take pain medication, but postsurgeries, postchemo, they gave me pain medicine from time to time, and I'd take the prescriptions up there, and they'd say, 'We don't know about that—that's from another doctor.' I'm like, 'Damn it, I go to seven doctors down there, OK?'" She eventually took her business elsewhere.

"You want to see the pharmacy?" Debbie asked, and I understood immediately she meant Sav-Rite, the shoebox of a building that had brought the town so much notoriety. I had driven past it a couple of times while on reporting assignments, and the parking lot was almost always empty.

"Let's go," said Debbie. "It's not far from here."

As we started to leave, my shoes scrunching dried leaves, I noticed a grave beside Timmy Dale's. It took me by surprise. It

251

was covered with dirt, just like Timmy Dale's, but there was no marker. Somehow I had missed it until now.

"Debbie, who's to the right?"

"My daddy loved this cemetery. He sat over here a lot." She wasn't answering the question. The drug companies had kept secrets. Now, she was keeping her own. The faint rumble of a freight train echoed across the river valley below.

"All right, tell me the story." I had my right hand shoved in my pants pocket, hiding the tremor that shook my index finger and thumb. It seemed to be getting worse by the day.

Debbie took three deep breaths and told it.

The cemetery workers had instructions to bury Timmy Dale beside Red Ed. They used a backhoe to dig the grave, the steel scoop clawing into the earth over and over, until a thud stopped them. It wasn't root, rock, or bone. "They get about three feet down, and they hit something," Debbie said. "Even back in the day, they didn't bury you three foot deep. They buried you six foot deep."

"What would you bury three feet deep?" For the briefest of moments, the question met with silence. In that void, in the stillness, I remembered her father's penchant for burying cash and other valuables, and his distrust of banks.

"Only things that you want to come back and get," she said. But those things, whatever they might be, were long gone, and Debbie was done talking.

We climbed back up the hill to the car, retracing the gravel lane out of the cemetery, then crossed the bridge that spanned the Tug Fork, turning into the Sav-Rite parking lot. The shadows lengthened, the sun fading behind the mountains, dusk falling. In the time we had left, we would strain to see through the darkness. There was more to unearth. We would keep digging.

Epilogue

The red light on my answering machine was blinking. I had just walked into my house after Parkinson's boxing therapy one afternoon in June 2019. I hadn't expected any calls. I pushed the play button. It was a familiar voice.

"Hi, this is Pat McGinley, calling for Eric. Just had a message that our federal-court-case opinion was issued. We won. Wanted to let you know that. You can reach me on my cell. Bye."

I called McGinley back immediately, unsure whether I had heard the message correctly. McGinley's tone was matter-of-fact, not celebratory. Turns out he was playing it cool. That was Pat.

The US Sixth Circuit Court of Appeals, he explained over the phone, had ruled that the Cleveland federal judge Dan Polster had exceeded his discretion by sealing the national DEA data that showed where all the opioid pain pills were flowing. The appeals court judges called Polster's decision "bizarre" and "irrational" and suggested the records were being used as a "bargaining chip" to spur a settlement between drug companies and the nearly two thousand towns, counties, and cities that had taken them to court.

Seven weeks earlier, I had accompanied McGinley and his wife, Suzanne Weise, to a federal courthouse in downtown Cincinnati. I was there to cover our appeal of Polster's ruling. An Associated Press reporter was the only other journalist in the courtroom. McGinley, along with *Washington Post* attorney Karen Lefton, stood before the three-judge panel, making the case that the public deserved the right to see the painkiller shipment database that the drug companies had fought so hard to conceal. A Justice Department lawyer representing the DEA countered that the release of the sales figures would hamper ongoing investigations of drug distributors and pharmacies. A lawyer for the drug industry, which had teamed up with the DEA to fight our request to release the data, contended that making the pill numbers public would be bad for business and allow competitors to steal customers. McGinley noted that the *Gazette-Mail* had already used a small part of the same database and published a flurry of stories about the companies' shipments to West Virginia in 2016 and 2017. The coverage, he said, didn't hamper any DEA investigations—in fact, it started a congressional inquiry—and the distributors continued to make billions of dollars on opioid sales. Two of the appellate judges seemed receptive to McGinley's and Lefton's arguments. One of them, Richard Allen Griffin, who was appointed by President George W. Bush in 2002, said it was troubling that the DEA would want to keep a lid on information "paid for by the taxpayers" that the "public has a right to know about."

By a 2–1 margin the court would rule in our favor. McGinley read parts of the decision to me over the phone. In strong, concise language, it directed Polster to clear the way for the release of the nationwide painkiller data.

By mid-July, he did just that. The *Post* was the first news outlet to get its hands on the database, the first to crunch the numbers, the first to publish a story. The *Gazette-Mail* received a Web link where we could download the disclosures, but the files were so big they triggered error messages on our decade-old desktop computers. Even after we figured out a way to open a few tranches, we didn't have a software program capable of analyzing them. Within hours, however, the *Post* had a blockbuster up on its website: The drug distributors had saturated America with 76 billion oxycodone and hydrocodone pills from 2006 to 2012. The records provided a road map to the painkiller epidemic nationwide, tracing the path of every prescription opioid manufactured and distributed. The states flooded with the most pills per person per year—West Virginia, Kentucky, South Carolina, and Tennessee—were also ravaged by a disproportionate number of prescription-overdose deaths. Rural counties were hit hard: Two in southern Virginia received the highest concentrations of pills per person, followed by Mingo County and nearby Perry County, Kentucky.

The following day, the *Post* made the DEA data publicly accessible in a searchable format on its website. Journalists across the nation downloaded it, publishing story after story about the number of pills that had inundated their communities and how the onslaught had fueled more and more deaths. "Avalanche of Opioids" read the headline in the *Montgomery (AL) Advertiser*. "Florida's Opioid Crisis: Billions of Pills, Millions in Campaign Cash" trumpeted the *Tampa Bay Times*. And the *Bangor Daily News* topped its coverage with "The Maine Pharmacy That Dispensed the Most Pain Pills." More than ninety articles were published over two weeks by reporters in thirty states.

The drug companies responded by blaming the DEA. The Feds had the same statistics. They should have stopped the shipments. Of course, the distributors reported those very numbers to the DEA. They knew all along the number of pills they were shipping. They knew precisely where their painkillers were going, year by year, town by town, pharmacy by pharmacy. They could have stanched the flow of opioids and slowed the deaths. Greed stood in the way. As overdoses surged, the companies shipped an increasing number of hydrocodone and oxycodone pills—from 8.4 billion in 2006 to 12.6 billion in 2012.

Two and a half years earlier, right after we published the explosive stories about the drug distributors' massive shipments to West Virginia, I had fielded dozens of phone calls from reporters, researchers, and grieving families who wanted to know how they could get the same data for their towns, cities, and states. I didn't have a quick and easy answer back then. I told them I was working on it; someday, I promised, the national pain pill numbers would get out. Now, they had, and the staggering details were helping the public understand the enormity of the epidemic. It had been a long journey into sunlight.

Acknowledgments

Debbie Preece wanted me to understand the origins of the opioid epidemic, a health crisis that had taken a deadly toll on America, her community, and her family. She guided me through the Appalachian mountains surrounding her hometown of Kermit, past coal mines long abandoned, under railroad trestles bleached by the sun, up hollows where people stored their garbage in cages to keep critters out. We ate slices of pizza and sipped lemonade from freshly squeezed lemons at Giovanni's restaurant. We toured the firehouse where a bronze plaque commemorates her father's service as chief. And we shared fears of the ruthless epidemic rotting our state, and of the ravages in our personal lives beyond our control and in spite of our good intentions. I'm grateful that she opened her home and heart to me and told her story unvarnished. Debbie has an extraordinary family. Her brother Tomahawk is the most dedicated firefighter I've ever known. Her sisters are pillars in the community. Her husband, David Dodd, is a gracious host.

I would never have met Debbie and her family without first crossing paths with Jim Cagle, who shared dozens of depositions and hundreds of pleadings. I spent many hours rooting through boxes in his office. His assistant, Melissa "Mo" Jackson, had an incredible knack for finding just the document I was looking for.

Jim also gave up hours and hours of his time answering my questions, many occasions over lunch. Rod Jackson, was equally helpful, taking my calls at night and on weekends. Rod always puts a smile on my face.

It's a safe bet that the distributors' shocking painkiller numbers would never have been made public without the tireless efforts of lawyers Pat McGinley and Suzanne Weise. They spent thousands of hours of their time researching case law, preparing and filing legal briefs, and arguing cases in court, but they didn't charge the *Gazette-Mail* a dime. They prevailed against enormous odds. They were aided by two other tremendous lawyers, Tim Conaway and Karen Lefton. Pat, Suzanne, Tim, and Karen prove that the sharpest legal minds don't always hang their hats at white-shoe law firms in New York and Washington.

At the newspaper, I've been fortunate to work with a cadre of journalists dedicated to "sustained outrage." They include Ken Ward Jr., Rob Byers, Greg Moore, Dawn Miller, Ben Fields, Lori Kersey, Scott Finn, Tara Tuckwiller, Alison Knezevich (Scott, Tara, and Alison were reporting on opioids years before I did), James Haught (an original "typewriter guerrilla"), Caity Coyne, Jake Zuckerman, Lacie Pierson, Ryan Quinn, Gary Harki, Dan Desrochers, Joel Ebert, Elaina Sauber, Craig Selby, Rosalie Earle, Patty Vandergrift Tompkins, Phil Kabler, and Andy Brown. (I suggested to Andy that we share the byline on the stories that would later win the Pulitzer, but he politely declined.) Also, thanks to Betty Chilton, Susan Chilton, and Trip Shumate for their long-standing commitment to investigative reporting in the face of enormous financial pressures. And to Doug Reynolds for rescuing the newspaper from bankruptcy and investing in journalism that makes a difference.

I didn't fully comprehend the enormity of drug distributors, such as McKesson, Cardinal Health, and Amerisource-Bergen until Michael Pryce-Jones contacted me in November 2016. Michael, an investment analyst with the Teamsters union, explained how the companies were profiting from pain, with total revenues of more than $400 billion a year. And he directed me to corporate filings that showed their CEOs' enormous salaries.

Nearly twenty-five years ago, I sent an unsolicited book manuscript to Nicole Aragi. It was about my experiences teaching English as a second language to Mexican and Guatemalan immigrants in a San Diego migrant encampment called the Ranch of the Devils. Nicole, who was just getting her feet wet as a literary agent in New York, took a chance on me and agreed to shop the manuscript. Alas, it never was published, through no fault of Nicole's. I had moved back East to be close to family and work at a sweatshop of a newspaper in Pennsylvania, leaving me with no time to make the changes Nicole had suggested, changes that would have made the manuscript publishable. Even so, Nicole sent me a baby gift (mini OshKosh overalls!) after my son was born a year later. I didn't forget the kind gesture. So when I decided to write this book and search for an agent, I contacted Nicole first. She was representing only novelists, but she recommended I talk to one of her colleagues who works with nonfiction authors. Frances Coady immediately recognized the story's potential and was an exceptional guide and mentor for this newbie to the publishing world. Frances found me an extraordinary editor, Kathy Belden, who embraced my ideas from the beginning and taught me how to take a series of events and shape them into a cohesive narrative. Her stellar team, including Sally Howe, Steve Boldt, and Lisa Rivlin also made important contributions.

In my battle with Parkinson's disease, exercise remains the best medicine. I'm blessed to have Jamie Tridico in my corner. Jamie leads two Rock Steady boxing classes and a Delay the Disease exercise class each week. She volunteers her time. She pushes us, motivates us. I'd rush home after classes, knowing it was my best time for writing. Exercise tends to lessen my tremor. Jamie's assistant coaches, Tori Charley and Ryan Fitzer, also put us through our paces. Tori stresses the importance of always keeping your

ACKNOWLEDGMENTS

guard up. Ryan straps on a protective vest and sparring mitts. He knows how to take a punch. I'm also grateful to the numerous volunteers, my fellow Parkinson's boxers, and to George Manahan, who runs the West Virginia Parkinson's Support Network. I have a whole new set of friends.

Other friends have helped me navigate the publishing world: Ken Armstrong, Chris McGreal, Mike Gordon, Jason Cherkis, Dale Maharidge, and Brent Cunningham. And I've appreciated the encouraging words from Nancy VanGilder; my mother, Bonnie; sister, Lori; Ry Rivard; and Marc and Amy Weintraub. I'm also grateful to Gracie "Sug" Wheelan, Matt Tolmach, and Brooke Ehrlich for recognizing the potential of this project beyond a book.

The manuscript's early readers—Amy Julia Harris, along with David, Buff, and Huck Gutman—kept me on track and provided valuable advice. Two additional readers, Joe Morris and Wade Livingston, went the extra mile. Joe helped edit first drafts, then polished later ones. Wade guided me through multiple revisions, helping me improve the flow of the narrative and find the exact words to tell the story best.

I'm not the first in the Eyre family to write a book. My grandfather, F. H. "Windy" Eyre, chronicled his travels in *Autobiography of a Forester*, published by the Society of American Foresters. Unfortunately, it's out of print, and I seem to have scooped up the last remaining copies. I can still picture my grandfather hunting and pecking on his portable typewriter. He died shortly after he finished the last chapter.

Now my son, Toby, watches me tap away on a laptop. Toby helped me fine-tune the manuscript. Toby and Morgan Hammock also shuttled me around Mingo County so I could conduct interviews with Debbie Preece and others.

Finally, my wife, Lori, realized the magnitude of this story from the beginning. At every step, she was there, providing unwavering support. No one has contributed more.

Notes

This book draws from interviews, documents, and data. No names were changed, no time lines altered. As a reporter for the *Charleston Gazette* and later the *Gazette-Mail*, I've written hundreds of stories about prescription drugs, pain clinics, overdose deaths, rogue doctors and pharmacists, and drug manufacturers and distributors—all of which shapes the contents of this book. I covered the courtroom battles over six years, following up by obtaining official transcripts of the proceedings. I filed scores of public records requests to pry loose tens of thousands of documents, as well as the opioid drug data that contained millions of transactions. I ventured to county courthouses across southern West Virginia and requested copies of lawsuits and legal briefs. And I interviewed hundreds of people about the opioid crisis— pharmacists, doctors, drug counselors, law enforcement officers, state legislators, lawyers, health statisticians and regulators, and families who lost loved ones to fatal overdoses.

Preface

Pharmacies' hydrocodone and oxycodone painkiller numbers came from the DEA's Automation of Reports and Consolidated Orders System (ARCOS) data sent by a DEA staffer to the West Virginia Attorney General's Office in January 2015. Additional pill numbers were revealed in a US House Energy and Commerce Committee report released in December 2018 (https://republicans-energycommerce.house.gov/wp-content /uploads/2018/12/Opioid-Distribution-Report-FinalREV.pdf).

Details about Sav-Rite Pharmacy's practices and its relationship to McKesson were drawn from the congressional report, depositions given by Jim Wooley in July and September 2008, and interviews with Jim Cagle between 2013 and 2019.

West Virginia attorney general Patrick Morrisey's ties to Cardinal Health and other drug distribution companies are disclosed in state campaign finance reports, federal lobbying disclosure forms, and a December 2015 report by the West Virginia Office of Disciplinary Counsel, which investigates complaints filed against lawyers.

1. A Death in Mud Lick

The recounting of William "Bull" Preece's death was compiled from interviews with Debbie Preece in Kermit and Mud Lick, depositions given by Debbie Preece and Jim Wooley as part of the wrongful-death lawsuit filed against Sav-Rite, William Preece's autopsy and workers' compensation reports, and prescription receipts and bottle labels recovered from his belongings.

The account of Dr. Donald Kiser's history was drawn from federal court filings related to his arrest, interviews with Debbie Preece and Cagle, Debbie's September 2008 deposition, the lawsuit filed against Sav-Rite, disciplinary records filed with the West Virginia Board of Osteopathic Medicine, and stories published in the *Charleston Gazette* and *Huntington Herald-Dispatch* from 2003 to 2008.

Debbie Preece recounted her trip to the state morgue and conversations with Dr. Sabet during her deposition. Her initial cooperation with Kiser was disclosed at her deposition and during interviews with me.

2. Prescription for Pain

Joe Rannazzisi was interviewed by me in July 2019. Details of the Distributors' Initiative were based on interviews with Rannazzisi and on the Cover2 Resources podcast (https://cover2.org/podcasts/), and the House Energy and Commerce Committee's report on opioid pill dumping in West Virginia. Additional information was gleaned from a December 2006 letter sent by the DEA to distributors.

The background on Purdue Pharma was based on articles published by the *Charleston Gazette* and the Associated Press from 2000 to 2007, and the "Prescription for Pain" series published in the *Lexington Herald-Leader* in 2003.

The email correspondence between McKesson employees about Sav-Rite's hydrocodone purchases was among supporting documents released by the House Energy and Commerce Committee (https://docs.house.gov/meetings/IF/IF02/20180508/108260/HHRG-115-IF02-20180508-SD005.pdf).

3. Kings of Kermit

Debbie Preece recounted her phone call about joining the Purdue Pharma lawsuit and her decision to contact Jim Cagle during interviews with me.

The history of the Preece family's drug-dealing past came from newspaper accounts in the *Charleston Gazette* and the *Charleston Daily Mail*, and from an April 1988 story headlined "Sweep of Hollows" published in the *Los Angeles Times* and written by Barry Bearak. Debbie also provided me with a VHS tape that included a 1988 interview with WSAZ-TV news reporter Kathy Brown. The broadcast took place just days before Debbie had to report to prison in Kentucky. The videotape also had a segment from ABC News about the federal raid and subsequent arrests.

Sav-Rite's motion to dismiss an attached expert witness/doctor's report were part of the wrongful-death lawsuit filed in Mingo County Circuit Court in Williamson. Kiser's letter from jail also was part of the court file. Debbie Preece and Jim Cagle disclosed Sav-Rite's initial settlement offer during interviews in 2017.

4. The Easter Bunny

Cagle provided me with a copy of the videotape taken by the private investigator in June 2008, along with the report he wrote afterward. Additional details were drawn from interviews with Cagle and "Tomahawk" Preece. I asked a law enforcement source to run a check on the license plate of one of the delivery trucks. It was registered to Miami-Luken, a company that delivered hydrocodone to Sav-Rite that year. The DEA's ARCOS data confirmed those shipments.

Wooley's depositions in July and September 2008 were part of the Mingo County court file. Additional information was drawn from interviews with Debbie Preece and Cagle.

5. Raided

The account of H. D. Smith's concerns about Sav-Rite's large hydroco-done purchases was part of the House Energy and Commerce Committee report, which released documents that chronicled H. D. Smith's site visit and interview with Wooley and his employee.

Details of the Sav-Rite raid were drawn from a search warrant affidavit filed by federal authorities in their criminal case against doctors and employees affiliated with the Justice Medical Complex. The *Huntington Herald-Dispatch* also reported on the raid in an April 2009 article headlined "Big Pill Network Exposed."

Sav-Rite customers' switch to Family Discount Pharmacy was spotlighted in the congressional report, which included emails and other supporting documents.

6. Addicts' Rights

This chapter drew on interviews with Debbie Preece and Jim Cagle, along with the lawsuits he filed in Mingo County Circuit Court and legal briefs filed in the cases after the pharmacists' and doctors' lawyers petitioned to have the West Virginia Supreme Court dismiss the lawsuits. The Supreme Court's opinion was published at http://www.courtswv .gov/supreme-court/docs/spring2015/14-0144.pdf. The two justices who cast dissenting votes, Menis Ketchum and Allen Loughry, were convicted on corruption charges in 2018. The charges were unrelated to the Mingo lawsuits. Loughry had previously written a book chronicling the history of political corruption in West Virginia.

7. A Step Up

Debbie Preece's pursuit of the delivery van was recounted in three separate interviews with her. DEA ARCOS data shows that Cardinal Health shipped drugs to pharmacies in Williamson, and in Inez and South Williamson, Kentucky.

Cagle provided me with a copy of Randy Ballengee's deposition. Details of Wooley's arrest and plea hearing were filed in the federal government's criminal case against Wooley.

The turnpike car chase was based on separate interviews with Cagle and Rod Jackson.

8. A Seismic Shift & 9. The Pekingese

These chapters drew from my notes, email correspondence with Morrisey, video, letters, and reports obtained from Morrisey's office in response to multiple requests filed under the West Virginia Freedom of Information Act, and stories I published in the *Charleston Gazette* in 2013. I also did follow-up interviews in 2017 and 2018 with sources who gave tips about Cardinal Health and provided the DVD of Morrisey's town hall meeting in Boone County. Morrisey held his inaugural ball on January 14, 2013, raising more than $36,000. Major contributors included Cardinal Health's political action committee, which gave $2,500 on February 19, 2013, and Capitol Counsel, the Washington, DC, lobbying firm owned by Morrisey's wife, which donated $5,000 to the inaugural fund on January 28, 2013.

10. Sustained Outrage

Some of the history of the *Charleston Gazette* was gleaned from a master's thesis, "Sustained Outrage: Owner/Publisher W. E. Ned Chilton and the *Charleston Gazette*, 1962–1987," by Edgar C. Simpson, while he was a student at Ohio University in November 2009. Also see Simpson's "Pressing the Press: William E. 'Ned' Chilton III's Investigation of Newspaper Owners," in the Winter 2011 issue of *Journalism History*. The *e-WV: West Virginia Encyclopedia* also was a helpful resource.

The newspaper's battle with the federal government over antitrust issues is based on my observations and stories published in the *Gazette*, *Daily Mail*, and *West Virginia Record*, and voluminous federal court documents filed in the case.

11. The Chase

The account of me questioning Dan Greear outside the Boone County Courthouse was based on my notes, a tape-recorded interview, and my story about the hearing published in the *Gazette* the following day.

I attended and reported on the courtroom hearing held by Judge King. I later obtained an official transcript of the hearing. Background details were drawn from the *Gazette*'s Freedom of Information Act lawsuit against Morrisey, which was filed in Kanawha County Circuit Court, and my coverage, published in the *Gazette*, of legal filings leading up to the hearing.

12. Hunger Games

Morrisey's subpoena was filed in Putnam County Circuit Court, along with a petition to enforce the subpoena.

The account of the *Gazette* merger with the *Daily Mail* was based on my observations, meetings I attended during the transition, and stories in both papers about it.

I obtained a copy of and reported on Judge King's ruling that Morrisey didn't have to comply with our FOIA request. The story was published in the *Gazette-Mail*. I attended the meeting with Rob Byers, Susan Shumate, and Pat McGinley at the *Gazette-Mail*'s newsroom office.

13. The Drop

I wrote about the document drop based on my observation of events. I kept a copy of the leaked email and envelope. To this day, I don't know who brought the document to my house.

Morrisey's threat of "sanctions" was based on emails sent by Tseytlin to McGinley, and interviews with McGinley and Anthony Martin, as well as emails between Martin and me.

Judge Stowers's rebuke of Morrisey's antitrust case against the *Gazette-Mail* was gleaned from coverage of the hearing by *Gazette-Mail* court reporter Kate White, along with a short discussion with White after the hearing.

14. Dragging and Lagging

The West Virginia Office of Disciplinary Counsel emailed me a copy of the report on the investigation into Morrisey's alleged conflicts with drug distributors. Morrisey's spokesman emailed me a written response.

My account of Morrisey's reluctance to sue McKesson was drawn from interviews with lawyers and administrators from both state agencies and the governor's office, along with emails they sent to Morrisey's office, and their public statements. Also, Jim Cagle speaks about this during a March 2016 deposition he gave as part of Randy Ballengee's unsuccessful defamation lawsuit against CBS News.

The court-approved agreement, called a protective order, to keep information confidential was on file in Boone County Circuit Court. The investigative subpoena was filed in Kanawha County Circuit Court.

15. A Door Cracked Open

The account of Judge Thompson's background and his initial decision to keep the state's lawsuit against distributors under seal was based on interviews with the judge and Jim Cagle. I gleaned Judge Thompson's decision-making process from the interview.

16. Eighteen Words

I attended and reported on the hearing to consider the *Gazette-Mail*'s motion to intervene and unseal in the AmerisourceBergen case.

I was on the conference call in which Emch asked that eighteen words from the unsealed lawsuit be redacted. I taped and transcribed the call.

17. A Legal Cartel

The testimony from Mike Smith, David Potters, and Mary Rochee was taken from their depositions given in 2016. Additional information was drawn from interviews with Potters, Jackson, and Cagle.

18. Home Court

I attended and reported on the court hearing to consider the *Gazette-Mail*'s request to unseal the state's lawsuit against Cardinal Health. Other information was gleaned from interviews with Conaway, court documents, and email messages between him and Henry Jernigan.

19. High Noon

I attended the "high noon" debate between Morrisey and Doug Reynolds. I also relied on *Gazette-Mail* reporter Andrew Brown's coverage of the event, an interview with Morrisey's campaign manager Kayla Berube, and a *Gazette-Mail* video recording of the debate, which is at https://www.youtube.com/watch?v=LMZCj52HBbM.

20. 780 Million Pills, 1,728 Deaths

This account of piecing together the story about the distributors' opioid shipments was drawn from interviews with Chelsea Carter, the Mullins family, and David Potters. The pill numbers came from the DEA's ARCOS data, while the overdose numbers were taken from the CDC's WONDER database and Center for National Health Statistics.

21. Misery's Price Tag

Two sources described the settlement talks with me during interviews in 2019. The details of the settlement were based on court documents and interviews with Judge Thompson, Tomblin, and his chief counsel, Peter Markham.

22. List of the Dead

I followed Kermit's mayor and town council members to the Mingo County Courthouse and interviewed Debbie and Tomahawk Preece at her home.

23. Damage Control

The accounts of the separate meetings with DEA administrators and AmerisourceBergen attorneys and executives were drawn from the interviews. I reported on Cardinal Health's legal brief that sought to assign blame for the addiction crisis to other parties. I interviewed two plaintiff's lawyers and Cardinal's spokeswoman about the legal strategy. The story was published in the *Gazette-Mail*.

A time line of the House Energy and Commerce Committee's investigation of pill dumping in West Virginia is at https://republicans-energycommerce.house.gov/opioids-pilldumping/.

24. A Death in Marrowbone Creek

The account of Timmy Dale Preece's overdose was drawn from interviews with Debbie and Tomahawk, as well as his obituary and video of the funeral parade.

A video of the October 2017 House hearing is at https://energycom merce.house.gov/committee-activity/hearings/hearing-on-federal -efforts-to-combat-the-opioid-crisis-a-status-update. I reported on the hearing for the *Gazette-Mail*.

25. "How in God's Name?"

A transcript of Mastandrea's interview with congressional investigators was provided to me by House Energy and Commerce Committee staff members, who also spoke to me later about his testimony. In July 2019, two former Miami-Luken executives (the company's president and chief compliance officer) were indicted on federal charges that they conspired to illegally distribute controlled substances. Miami-Luken, as a company, was also charged with the same crime.

The press conference about the DEA's inaction, which I covered, was livestreamed, but only a portion was recorded at https://www.facebook .com/energyandcommerce/videos/highlights-from-ecs-press-conference -on-opioid-crisis/10155758373001311/.

26. Bankrupt

The account of the *Gazette-Mail*'s bankruptcy was drawn from office emails, bankruptcy filings, and stories in the *Gazette-Mail*, *New York Times* (https:// www.nytimes.com/2018/02/02/business/media/west-virginia-newspaper -charleston.html), and *Washington Post* (https://www.washingtonpost.com /business/economy/a-west-virginia-newspaper-is-in-bankruptcy-the -powerful-coal-industry-celebrates/2018/02/16/f0e3d4e4-085c-11e8-8777 -2a059f168dd2_story.html).

27. Paper and Tapes

I reported on the congressional hearing during which panel members questioned Patterson about the DEA's role in combating the opioid crisis. My story was published in the *Gazette-Mail*. Additional information and a recording of the hearing are at https://republicans-energycommerce .house.gov/hearings/drug-enforcement-administrations-role -combating-opioid-epidemic/.

A source provided me with a copy of the email Denise Morrisey sent to McKinley's office. A transcript of the hearing is at https://www.govinfo

.gov/content/pkg/CHRG-112hhrg80861/html/CHRG-112hhrg80861
.htm. A video recording of the hearing is at https://www.youtube.com
/watch?feature=player_embedded&v=JqQH1kUDdn0. Additional information was drawn from an interview with Joe Rannazzisi.

28. "What's the Punishment?"

I reported on the hearing for the *Gazette-Mail*. More information,
including a video recording and transcript, is at https://republicans
-energycommerce.house.gov/hearings/combating-the-opioid-epidemic
-examining-concerns-about-distribution-and-diversion/.

29. Whose Pills?

The account of the *Gazette-Mail*'s attempt to unseal the DEA ARCOS
data for the entire nation was based on court documents filed in northern
Ohio federal court, as well as conversations and interviews with McGinley, Sauber, and Lefton.

30. Three Feet Deep

I twice accompanied Debbie to the cemetery in 2018. This account was
based on my first visit.

Epilogue

I made the trip to Cincinnati to report on the appeals court hearing for
the *Gazette-Mail*. I also wrote stories about the court's decision to over-
rule Polster and unseal the ARCOS data. Comprehensive coverage of
the release of the data and its aftermath can be found on the *Washington
Post* website at https://www.washingtonpost.com/national/2019/07/20
/opioid-files/.

INDEX

INDEX

INDEX

About the Author

Eric Eyre has been a reporter in West Virginia since 1998. In 2017, his investigation into massive shipments of opioids to West Virginia's southern coalfields was awarded a Pulitzer Prize. He lives in Charleston with his wife and son.